AMERICA'S FUTURE IN TOXIC WASTE MANAGEMENT

Recent Titles from Quorum Books

Entrepreneurial Science: New Links Between Corporations, Universities, and Government
Robert F. Johnston and Christopher G. Edwards

Creating Investor Demand for Company Stock: A Guide for Financial Managers
Richard M. Altman

Interpreting the Money Supply: Human and Institutional Factors
Loretta Graziano

Business Demography: A Guide and Reference for Business Planners and Marketers
Louis G. Pol

International Space Policy: Legal, Economic, and Strategic Options for the Twentieth Century and Beyond
Daniel S. Papp and John R. McIntyre, editors

Inquiry and Accounting: Alternate Methods and Research Perspectives
Ahmed Belkaoui

A Problem-Finding Approach to Effective Corporate Planning
Robert J. Thierauf

Managing Risk in Mortgage Portfolios
Alex O. Williams

Evaluating Employee Training Programs: A Research-Based Guide for Human Resources Managers
Elizabeth M. Hawthorne

Current Budgeting Practices in U.S. Industry: The State of the Art
Srinivasan Umapathy

Developing Negotiation Skills in Sales Personnel: A Guide to Price Realization for Sales Managers and Sales Trainers
David A. Stumm

Training in the Automated Office: A Decision-Maker's Guide to Systems Planning and Implementation
Randy J. Goldfield

The Strategist CEO: How Visionary Executives Build Organizations
Michel Robert

America's Future in Toxic Waste Management_____

LESSONS
FROM
EUROPE

Bruce W. Piasecki and Gary A. Davis

Foreword by
LYNTON K. CALDWELL

Q

QUORUM BOOKS
New York • Westport, Connecticut • London

363.728
P 57a
146838
Jac, 1989

Library of Congress Cataloging-in-Publication Data

Piasecki, Bruce.
  America's future in toxic waste management.

  Bibliography: p.
  Includes index.
  1. Hazardous wastes—Europe.   2. Hazardous
wastes—United States.   I. Davis, Gary A.
(Gary Allen), 1952-   II. Title.
TD811.5.P523   1987   363.7'28   87-2559
ISBN 0–89930–113–4 (lib. bdg. : alk. paper)

British Library Cataloguing in Publication Data is available.

Library of Congress Catalog Card Number: 87–2559
ISBN 0–89930–113–4

First published in 1987 by Quorum Books

Greenwood Press, Inc.
88 Post Road West, Westport, Connecticut 06881

Printed in the United States of America

∞

The paper used in this book complies with the
Permanent Paper Standard issued by the National
Information Standards Organization (Z39.48-1984).

10 9 8 7 6 5 4 3 2

Contents

Illustrations ─────────────────────────────

Foreword

LYNTON K. CALDWELL

America's Future in Toxic Waste Management provides a critical cross-cultural study of hazardous waste management issues. The timeliness of this international study should require no further comment, but most Americans are not sufficiently informed about the growing menace of toxic contaminants in the environment. Some see the toxics crisis as primarily a local issue and remain unaware of its now-global dimensions. Few are aware of the dilemmas that this problem presents to corporate managers and government leaders. Still fewer understand the sustained effort and costs that will be required to initiate a superior response.

Piasecki and Davis understand these restrictions, yet they add to the American scene a bright new insight: safer alternatives to land disposal already exist in Europe. While America continues to dump the bulk of its worst toxic wastes, these European solutions remain unstudied except for the work of a few technical experts. American society has moved into a high-tech, large–waste-volume era with low-tech assumptions and dated political institutions. America's toxic waste problem, the authors argue, is a consequence of this insufficient policy response to the swift and ubiquitous developments in industrial chemistry across the last 40 years.

This book shows the direction that must be taken if America's toxic waste problem is to be resolved. There is great resistance, especially in America, to

A major figure behind the passage of the National Environmental Policy Act (NEPA) and the establishment of the environmental impact statement (EIS) process, Lynton Caldwell is author of *International Environmental Policy* the reordering of priorities, the reallocating of resources, and the restructuring of political and corporate institutions that effective waste management requires. Before any of the aforementioned changes can occur, however, citizens must come to understand the scope of the dilemma and the diversity of problems arising from it. They must be shown examples of programs that successfully shift the burden of waste management off the land. Moreover, political representatives must be provided with workable evidence of the next steps a nation can take after land disposal. *America's Future in Toxic Waste Management* answers these needs in cogent and effective prose.

Copies of the book, therefore, should be in every city hall, county courthouse, and public library. It should be read by civic leaders as well as public officials, and it should be required reading in civil and sanitary engineering, political science, and environmental studies courses. We can no longer afford the uphill battle against leaking landfills and other land disposal methods such as the still-popular technique of deep-well injection. As Piasecki argued convincingly in *Beyond Dumping* (Quorum Books, 1984), waste reduction, recycling, and treatment are ideas whose time has come. This new book explores the specific means by which Americans can now secure their use. Moreover, as the first book to examine European legal, technical, and planning initiatives to progress beyond land disposal, it is a major contribution toward conceptualizing the role governments can play in pursuing superior waste management practices in a rational and informed manner.

THE INFLUENCE OF THE AMERICAN SOCIAL CHARACTER

In seeking an effective strategy, it makes sense first to examine why and how the problem arose. This is not the common approach to problem solving in American politics, where the customary question is not *why* but *who*. Who is responsible? Who is to blame? Who should be punished? Who should pay? Who should take corrective action? This same knee-jerk response accounts, in part, for the continued failure of recent efforts—often sincerely undertaken by elected officials—to deal effectively with the regulation of hazardous wastes. Piasecki and Davis resist this approach, matching their recommendations with what they term "the intrinsic properties of waste management."

Moreover, the disastrous mismanagement of toxic waste in America is, the authors suggest, neither a technical nor a financial problem at its core. Inadequate and inappropriate legal doctrines have permitted careless, wasteful, and harmful land-based waste disposal practices. But the principal causes for policy failures are rooted in American beliefs about its resources and the role of government in improving the management of those resources. Thus the authors of this book examine not only the roles European governments have assumed to end dependence on land disposal but also explore the thought processes underpinning the focus on toxic wastes as resources.

Perhaps the most significant single trait of the American social character is its complacent and self-serving devotion to individual needs rather than to goals for the entire society. Americans remain a people defined by a vast geography and an optimistic and reaffirming political mythology. This unduly narrow emphasis on individualism and private property rights may have kept Americans from assuming the leadership roles government can play in protecting society from the hazards of toxic pollutants. While Americans pride themselves on their massive public works projects in transportation, irrigation, and entertainment, they have neglected the public management of waste as a politically appropriate activity.

Circumstances of the American frontier placed a survival value on cooperation, and citizens of diverse backgrounds continue to this day to rally on behalf of perceived common causes. But a spirit of civic cooperation is not the same as a commitment to perpetuating the social and environmental character of communities. Some religious groups, such as the Amish and the Mormons, have occasionally shown such commitment; more have regarded the earth and its environment as a temporary place of residence and have placed no great value on its natural conditions.

Added to the subliminal bias that favors individualistic, short-term decision making has been the mobility of the American people. There has been little incentive for many Americans to take interest in long-term community planning, let alone the sophisticated waste collection and management systems the authors assess in Europe. There is even less incentive for them to support tax assessments and bond issues to pay for public works the benefits of which they do not expect to reap. This lack of long-term environmental concern has been reinforced by a perceived abundance of cheap land and fresh water—conditions of special relevance to waste disposal practices. In reconstructing the logic by which European governments created an infrastructure for superior waste management, this book highlights the costs and shortcomings inherent in these American traditions.

REQUIREMENTS FOR SUPERIOR WASTE MANAGEMENT IN AMERICA

The toxic waste crisis is the critical acid test of the more general problem of waste management. Both aspects, general and specific, have become extremely problematic in our time. Great increases in the numbers and concentrations of population only partially account for these problems. Unfortunately, American society has not matched its production-oriented innovations with concepts or institutions appropriate for management of its by-products. Indeed there is widespread belief that technical innovation requires no governmental management, and the idea of subjecting it to industrial policies or social control is perceived as mischievous, dangerous, or downright harmful.

A consequence of the sum of these social conditions has been the unpreparedness of American society to cope with the residual products of industrial chemistry and the immense volume of toxic and other solid, liquid, and gaseous wastes.

This brief commentary should help one understand why it is much less difficult to rally community support for a new athletic complex or art center than for a hazardous waste treatment facility. Support for environmental amenities does not come easily in most American communities; securing a consensus for closing down the "low-cost" county dump and building a "high-cost" detoxification facility is even more difficult.

This book confronts these problems directly, explaining how an alternative route works. Recognizing American restraints, it provides ample evidence that public attitudes can change rapidly under the stress of threatening events. Eventually the arguments set forth here must move the American government to systematic action that resolves rather than merely alleviates the waste problem.

This book can serve as a foundation stone for the construction of this comprehensive waste management policy in America. The examples of successful systems described herein are especially important, since opportunities for similar changes are suddenly arising in America. The new Resource Conservation and Recovery Act amendments, for example, are likely to add urgency to the already-heated debates over waste reduction, recycling and treatment for at least the next ten to thirty years. *America's Future in Toxic Waste Management* will further fuel those debates, since even knowledge of the mistakes of the Europeans, the authors note, should prove as helpful as knowledge of their successes. The following description and analysis of the European experience provides a record to which Americans may refer as they reconsider their own policies regarding waste management. My only hope is that the next steps America takes prove accessible not only to the specialist in environmental management, but to the American public at large. This lucid book goes far in fulfilling that hope.

Acknowledgments _____

Doing business and research in foreign nations can be both frustrating and fascinating. The lost mail, the late trains, the mistaken messages confuse and amaze. Yet the net result captures a sense of wonder. Suddenly, a new vista is before you, new trains of thought arrive that reach beyond one's own culture. Such was the case throughout the four years we worked on this book.

We'd like to thank the thousands of people who gave us the right directions. Without these simple signals, the project would have been far more complex. We'd also like to thank the hundreds of professionals—lawyers, engineers, policy analysts, government officials, and business leaders—who shared with us their advice and insights. In this department, our appreciation for the help of Marianne Ginsburg of the German Marshall Fund, Ron Stevens of the Joyce Foundation, and Bill Colglazier of the University of Tennessee stands out. Moreover, we'd like to thank the many European officials, experts and translators who made this nine-nation study both possible and enjoyable, particularly Benno Risch, Dr. Hans Sutter, Dr. Brian Wynne, and Dr. Joanne Linnerooth.

Ultimately, research is only as strong as the institutions that fund and house the effort. In this light, we'd like to thank Cornell University's Center for Environmental Research, the University of Tennessee's Waste Management Research and Management Institute and the progressive engineering setting of Clarkson University, where both great books and hard facts are taught. Our funding came, in a timely fashion, from the German Marshall Fund, the Joyce Foundation, the Department of Energy's National Hazard Chemical Waste Program, the Congressional Office of Technology Assessment and the International Institute for Applied Systems Analysis in Laxenbourg, Austria.

Andrea Masters and Janie Wilson must also be thanked for their productive patience. Thanks, also, to Elizabeth Richert, Robin McClellan, and the students at Clarkson for their help as research assistants or astute questioners.

AMERICA'S FUTURE IN TOXIC WASTE MANAGEMENT

1

Restructuring Toxic Waste Controls: Intrinsic Difficulties and Historical Trends

BRUCE PIASECKI AND GARY DAVIS

While debate rages about what to do with the 200 million tons of hazardous chemical wastes generated in the United States each year, several industrialized nations in Europe are quietly reducing their toxic waste burdens. The hazardous waste systems in these countries resemble chemical plants rather than the leaking landfills that litter America. Their successful operation indicates that our uphill battle to clean and contain thousands of leaking toxic dumps, pits, ponds, and industrial lagoons represents an expensive trip down the wrong road.

Over fifteen years ago land-conscious Europeans decided that dumping chemical garbage on or into the ground was neither safe nor economical. Instead, several countries developed policy responses designed to reduce the amount of toxic wastes generated at the source and use modern chemical technologies and high-temperature incinerators to further minimize the need for dumping. Although most forms of waste treatment produce small quantities of residues that must be managed, the net result is that the risk, waste volume, and overall costs to both society and its leading firms are greatly reduced.

In 1973 Denmark established the Kommunekemi facility in Nyborg, which now destroys over 70 percent of the nation's hazardous wastes. This integrated treatment facility also recovers heat from its toxic waste incinerators to supply

The authors wish to thank Dr. Brian Wynne, whose provocative work provides the conceptual groundwork for the inquiry into intrinsic properties examined throughout this chapter. See Brian Wynne, *Risk Management and Hazardous Waste: Implementation and the Dialectics of Credibility*, Springer Verlag, New York, 1988.

Nyborg's 18,000 residents with 35 percent of their heating needs. Five years earlier, local governments within the West German state of Bavaria established the earliest hazardous waste treatment center in the world. With almost twenty years of operating experience, Bavaria's ZVSMM facility contains treatment components many Americans now hope to site and finance. Other European governments, including those of Sweden, Austria, Finland, and the Netherlands, are following these early examples by minimizing access to dumping and requiring the use of recycling, incineration, and treatment technologies. Part 1 of this book explores at length the role technology has played in Europe in shifting the burden off the land.

But the European approach involves far more than sophisticated technology. Some governments in the Old World realize that reliable new techniques are essential for improved management, but they also recognize the need for comprehensive public policies and consistent government involvement to introduce and use superior techniques appropriately and cost-effectively. The result, in several European countries, has been a creative combination of private and public enterprise that manages toxic wastes with a minimum of expense and red tape. Part 2 explores the competing goals governments must balance to achieve these effects.

A willingness to conceive of the hazardous waste problem in this context of public-private joint venture may prove the most important legacy of the European example, helping Americans see beyond their inherited institutional and ideological blinders. In response to the intrinsic challenges of waste management, the various European approaches, while different in detail, have three pivotal factors in common:

1. A willingness to stabilize a waste treatment market through a fundamental policy change rather than through a series of small unrelated financial incentives and regulatory refinements that encourage incremental changes in private business practice

2. A knack for cooperative management and joint financing between regional, state, and federal governments and groups of private industries and concerned citizens

3. A profound distrust of land disposal, despite engineering advances in constructing and monitoring landfills, deep wells, and industrial lagoons

These factors clash with some longstanding American biases. Perhaps the most crucial is the prejudice that still favors upgrading existing land disposal methods such as the recent push for expanding the use of deep-well injection in the United States. In the decade since Love Canal, the United States has poured billions of dollars into patching up dumps and improving land disposal techniques. In a willful defiance of thermodynamics and common sense, we have built up our landfills with precipitation caps, double liners, restraining walls, and collection systems for leaching wastes. Despite these technical reforms, the recent reauthorization of the Resource Conservation and Recovery Act (RCRA, November 8, 1984) aggressively expands the boundaries of the American response with an almost total redirection of priorities away from land disposal toward waste reduction, recycling, and treatment. To fulfill the spirit of this law, Americans must now assess the European experience.

The United States has had its own successes in establishing large-scale public environmental management projects. Faced with massive flooding, deteriorating urban water supplies, and municipal sewage, this country evolved over its first 200 years a cooperative public works program to better manage these problems. Federal, state, and local governments coordinated the construction of an infrastructure that includes dams, water mains, sewers, and a world-renowned highway network. Moreover, within the past 75 years, the agricultural extension service has been developed as a joint public-private venture to maintain and revitalize our farmlands. Both programs offer robust American models for responding to the ubiquitous environmental threats inherent in toxics contamination. Yet to date little has been done along these lines involving toxic wastes.

While European strategies cannot simply be transferred wholesale within the U.S. free-enterprise system, certain components of the European programs are ripe for adoption. Part 3 explores those areas most ready for U.S. adaptation— siting initiatives and collection schemes—and reflects on how such applications might further build the infrastructure America needs to move beyond land disposal. *America's Future in Toxic Waste Management* outlines a new home for hazardous waste policy, calling for a fundamental restructuring of our response. The design of this home is transatlantic: it combines the European initiatives in creating a treatment infrastructure with recent American tactics for defining and regulating risk. This deliberate merger of the infrastructure and risk management approaches combines the advantages of both traditions. How this restructured response better matches the intrinsic properties of the problem is the subject of this book.

WHAT THE OLD RESPONSE LEFT OUT: TRADITIONAL EXCLUSIONS IN HAZARDOUS WASTE POLICIES

Policy responses to hazardous waste issues are neither innate nor fixed. They are socially defined by policymakers, often in spite of the prominent technical dimensions of the issue. The current American definitions of hazardous waste issues show how unnatural and even arbitrary such boundaries may be. Although these boundaries can be explained by noting the historical forces that shaped the response, they remain inadequate whenever they do not address the intrinsic characteristics of waste problems. Once set, they have had profound effects upon the orientation of American policymakers and, consequently, upon the policy options that are considered as solutions to the problem. The result is a highly controversial hazardous waste policy most notable for its exemptions and exclusions. What follows highlights four of these still-dominant boundaries and suggests the dangers implicit in such exclusionary tactics.

Exclusion from the Universe of Hazardous Chemicals

The hazardous waste problem has been kept separate from the much larger hazardous chemical problem. There are approximately 80,000 chemicals in commercial use and approximately 1,000 new ones introduced each year. Although most of these are not hazardous, only a small number have been thoroughly tested for their toxicity. An even smaller number shown to be toxic are regulated effectively, since their production, distribution, use, and disposal still involve serious impacts. Focusing regulatory attention on a shortened high-visibility list of chemicals in hazardous waste creates a myopic perspective that today still dominates American hazardous waste policymaking.

A further complication is born of this artificial separation: how does one define the point at which a hazardous chemical becomes a hazardous "waste"? This definitional requirement has proven problematic in the United States, especially concerning reuse and recycling ventures. Changing economics often legitimately make yesterday's waste today's raw material, and most nations are seeking to encourage such recycling. But this fuzzy demarcation between waste and reuseable by-product has been frequently exploited in the United States by unscrupulous parties. The roles European governments play to respond to this structural paradox are discussed at length in Chapter 6, where West Germany's and France's innovative programs to make waste oil recoverable are examined.

Exclusion from Traditional Media-Specific Laws

Hazardous waste is usually dealt with separately from air pollution and water pollution despite the links among the three. One of the reasons America suffers so extraordinarily from leaking dumpsites is that air and water pollution regulations were tightened in the 1970s without parallel requirements for placement of the toxic residues removed from these media. Recognizing this problem of cross-media transfers, Europeans encouraged waste reduction at the source through sustained government intervention. Chapter 2 examines the institutional means by which these reduction strategies were established.

Exclusion of Household Toxics

Hazardous waste subject to regulation and control generally does not include hazardous chemicals discarded by households, although hundreds of commercial products contain substances that can cause serious public health and environmental problems when disposed of improperly. Household wastes are not excluded on technical grounds, but on administrative and political grounds. Chapter 10 examines at length the problems created by household toxics, evaluating recent collection programs in Europe and select American states.

Exclusion of Related Waste Streams

Despite the high toxic metal content of some sewage sludge and mine and mill tailings, these waste streams are often excluded from hazardous waste regulatory programs. These and similar exclusions are persistently legitimated in terms of administrative pragmatism despite their public hazards. By examining the value of integrated toxics management, our concluding remarks underscore the need for a consistent regulation of all hazardous materials.

The net result of these common exclusions is severe: the universe of regulated substances in America is both truncated and distorted. Without a comprehensive response the demand for hazardous waste treatment is weakened, delaying technical innovations and new facility construction. It may well prove overwhelming for Americans to attempt to control all hazardous materials by revising existing hazardous waste regulations. Moreover, this patchwork of reforms might only increase the distortions. In an effort to find a fresh way out of these dilemmas, we turn now to a brief review of the historical forces that created and then deepened these original exclusions in the United States.

WHY THE OLD RESPONSE STILL DOMINATES: HISTORICAL TRENDS AND INSTITUTIONAL ENTRENCHMENT

As America reacts to one hazardous waste disaster after another, it becomes more and more clear that our exclusionary approaches are not working. The reasons for this failure are complex, yet three pressure points keep crumbling American programs. Regulatory programs have evolved in an entirely reactive fashion, building piecemeal on old assumptions about solid waste that proved insufficient in the light of technical discoveries emerging almost daily. Second, as the administrative roots of the response took hold within the context of ordinary garbage disposal, a long love affair with land disposal resulted. This dependence on land disposal, which demanded concentrated budgetary, regulatory, and remediation effort, has dominated the design of policies in America and eroded consideration of more effective treatment alternatives. Finally, the sheer complexity of the hazardous waste predicament, which is only a subset of the larger hazardous chemical problem, has intimidated many American policymakers into clinging to the original edifice. Thus exclusionary responses continue to place boundaries around the problem that artificially separate it from a more integrated approach. It is as if America's traditional policy response constituted a building where the ceiling, entrance, and back wall kept collapsing.

In the early to mid-1970s, the regulatory concern over toxics arose directly from changes in ordinary garbage collection and disposal. Prior to that time, most waste materials from industry that could not be discharged into rivers and streams were either dumped into municipal landfills with household garbage or into unlined landfills at the generator's industrial sites. Early concern over hazardous waste grew out of a recognition that some wastes being disposed of with ordinary garbage were dangerously flammable, reactive, corrosive, or toxic.

Another historical oddity involves the isolated status the problem was given at its outset. Hazardous waste regulation did not evolve with an explicit mission to protect groundwater or any other specific environmental medium. The regulatory response was therefore different—and separate—from that of the more established programs controlling air and water pollution. America's primary regulatory response there was directed toward short-term managerial considerations, which also predated concerns over the health and environmental impacts of the growing commerce in hazardous chemicals. Thus there was no attempt to fit hazardous waste into the broader hazardous chemicals issue.

When local governments began to recognize the problem in the early 1970s, they did not evolve prompt responses. Instead, since expertise and resources were generally scarce at the local level, pressure gradually arose for the centralization of government controls. Even centralization lagged until the late 1970s, when major disasters like Love Canal made hazardous waste a pressing public issue. Its rapid rise into national visibility further fueled the demand for centralized control, and the federal government's frantic response partially overlooked RCRA's initial emphasis on resource recovery and conservation. The result further distorted the problem, with EPA's programs disproportionately preoccupied with land disposal.

The systems of central control that came out of the push for stronger regulation of hazardous waste did not change the initial exclusionary boundaries adopted earlier at the local level. In fact, exclusions became more entrenched as a result of central control, since administrative difficulties became magnified at this level. The United States' small generator exemption, which was part of the RCRA regulations for the first four years of their implementation, is a prime example.

The evolution of hazardous waste management from garbage collection also helps explain America's peculiarly long reliance on landfill disposal. Both the practitioners and regulators had long experience with landfill disposal; thus both parties were slow to address hazardous waste management as a chemical engineering problem instead of a dirt-moving one.

Today Americans are in a second wave of public debate on hazardous waste issues. It is the purpose of this book to inform this debate. What follows recounts, in composite form, the logic by which European policymakers decided to deconstruct some of these old exclusionary boundaries and begin formulating a more comprehensive response. We have stated this logic abstractly, both in an effort to avoid the recital of countless names and details (for which please see Appendix 1) and to display the intellectual challenge inherent in restructuring a more workable response in America.

BUILDING A NEW RESPONSE

Several intrinsic properties of hazardous waste set it apart from other environmental issues. These properties produce vastly different regulatory results in differing political and institutional settings, but this book argues that all must be accommodated by any nation that seriously confronts its hazardous waste problems.

Hazardous Waste and Hazardous Waste Generators Are Extremely Diverse

Since hazardous chemicals are ubiquitous in modern industries, wastes containing hazardous components are generated by nearly all industrial and commercial activities. Europeans have based their response on this sound expectation. In the United States, even with exemptions for household waste, mine tailings, and other potentially hazardous wastes, there are approximately 66,000 hazardous waste generators, excluding recently regulated small quantity generators, and about 50,000 hazardous waste shipments each year.[1] On both sides of the Atlantic generators range in size from multinational giants such as Dow Chemical or Hoescht Chemical (in the Federal Republic of Germany) to small family-owned paint shops and metal-plating firms. Moreover, rapidly emerging industries like microelectronics are characterized by small firms, yet generate highly toxic residues. The methods for management of hazardous wastes are also many and diverse, ranging from simple land disposal to sophisticated transportable treatment units (TTUs). Furthermore, the management of wastes can be performed either at the site of generation or at a facility located somewhere hundreds of miles from the point of generation.

This extreme variety within the hazardous waste arena offers a striking contrast to the nuclear waste industry. Nuclear waste has been regulated separately from hazardous waste, as it is a fairly narrow range of well-characterized radionuclides in relatively narrow bands of composition. Nuclear waste is produced by a fairly small number of generators and in much smaller quantities than hazardous waste. Furthermore, the nuclear industry itself is monolithic, with few centers of decision, control, and responsibility. Thus the extreme diversity of hazardous waste creates the difficulties in defining the scope of the problem and makes regulatory exclusions so commonplace and appealing. Whereas nuclear waste regulators generally know how much waste is produced and where it is going, this is not the case with most hazardous waste regulators despite earnest attempts to gather and evaluate this information.

The phenomenal differences in size among hazardous waste generators also prove puzzling for policymakers. For example, an exemption for small generators was granted in the United States for pragmatic administrative reasons, but indications that small generators are responsible for significant quantities of hazardous waste that end up in municipal landfills led Congress to lower the exemption by a factor of ten in 1984. The response of small firms to regulatory controls is often different from that of large firms. Small firms tend to have less control over their economic and social environment and are thereby sometimes less concerned with the long-term effects of their decisions. Innovations in

hazardous waste management may be neglected since such firms also tend to
lack expertise and resources to keep abreast of, understand, and implement
complex regulatory schemes.

Large firms, on the other hand, generally occupy a more secure economic
position and have more control over their immediate economic and social environment. As a result, they are likely to have a longer-term decision-making
horizon than small generators. They also have a national or international image
to protect and usually the expertise and resources to keep abreast of and comply
with environmental requirements. Thus larger firms tend to react differently to
hazardous waste regulations and, despite being the generators of most hazardous
waste, may actually prove to be a smaller part of the compliance problem than
is often assumed in the United States.

An irrepressible issue raised by this extreme variety of waste generators is
whether there is, in fact, a uniform field in which government can plan and act.
America's recurrent and persistent assumption of uniformity where it does not
exist has created severe pitfalls for its regulatory schemes. Thus far the result
has been regulations that are too stringent on one end of the hazard scale and
not stringent enough on the other, seriously affecting their credibility and effectiveness. The assumption of uniformity has also resulted in overly complex
regulations, which must be designed to take into account a boggling range of
possibilities. As a consequence, American agencies encounter severe problems
in securing the necessary expertise and experience to adequately control the
richly diversified regulated community.[2]

Lack of Professionalism in the Waste Management Sector

There has been a historical lack of professionalism in the field, largely due
to the evolution of hazardous waste management from the arena of household
garbage at the local level. There was no professional status attached to garbage
disposal and no systematic formal attention or resources committed to it. As a
result, the dangers of land disposal of hazardous waste went unrecognized for
too many years. Moreover, the need for more effective regulations far outstripped
the ability of the initial cadre of solid waste specialists to deal with the complex
problems of risk management for hazardous chemicals.

In recent years the level of professionalism has increased dramatically, but
most countries still lack the large numbers of experts needed to meet the demands
of industry and the regulatory agencies. Part of this failure can be attributed to
the time lag necessary to retool the educational system. Another source of delay
is that the field requires many different types of technical expertise in a multidisciplinary approach that also combines legal, business, and communication
skills. Thus it is difficult to identify a single professional discipline in the educational system to build upon.

This lack of professionalism is still a burning issue in America at the state
and local level in government and at the plant operator level in industry (par-

ticularly for smaller firms) where most of the burden of implementing hazardous waste policies falls. Past managerial inadequacies have led to a profound public distrust of hazardous waste regulators and of the hazardous waste industry at large and are even partially responsible for debilitating some needed and reliable reforms. Partly because of this entrenched lack of professionalism, the European nations discussed in the course of this book determined that governments should play an increased role in the siting, financing, and operating of hazardous waste facilities. Part 2 of this book evaluates the role of governments in controlling the risks of hazardous waste.

Hazardous Waste, If Sent Off-Site, Is a Packaged Waste

Since generators of hazardous waste have the option of sending their waste somewhere else for management, managing the risks of hazardous waste is fundamentally different from controlling routine pollution emissions. Hazardous waste is first packaged, then shipped in various containers by many modes of transportation. The life cycle of these packaged wastes in a free market can undergo many stages, including the following:

1. In-plant process generation of the waste
2. Mixing with other in-plant by-product or waste streams
3. Packaging in drums, tank trucks, or other containers
4. On-site storage: long-term or short-term
5. Collection and initial transportation
6. Interim storage: commingled or stored separately
7. Mixing, repackaging, retransportation
8. Initial processing (recycling, pretreatment, blending)
9. Repackaging and retransportation
10. Final processing (incineration, treatment)
11. Disposal of residues (landfill, retrievable storage)

Since the U.S. approach often involves many different firms along the way, the prospects of mismanagement are significantly magnified throughout. Although many packaged wastes do not go through all of these stages, the above life cycle is not atypical. Through these stages a packaged waste may undergo several changes in composition, physical state, economic value, ownership, and control.

In contrast to conventional modes of pollution control, a packaged waste is not dispersed into the environment in any predictable manner. Since air pollution and waste water discharges are mostly emitted from discrete sources, their risks can be determined by a knowledge of the substances emitted, their pathways to public exposure, and their health effects. Conventional ambient standards and emission limits can be set to control these risks. Although there are scientific uncertainties in each of these steps, such risk assessments and risk management decisions are now an accepted part of environmental regulation. Packaged hazardous waste, however, changes hands many times and undergoes several chemical and physical transformations before it is dispersed into the environment. This delayed entry into the environment often occurs at a location far from the site of generation.

The final dispersal of the waste into the environment is still subject to the same scientific uncertainties as conventional pollution, but a significant extra dimension of uncertainty is added by the complex chain of events that take the waste from the generator to its final dispersal. For instance, because hazardous waste is not immediately dispersed, there is always an ambiguity about whether it will be used as a resource (recycled). As a result, a regulatory system that only regulates the ultimate processing of waste will overlook material that is used as a resource even if this phase of reuse creates similar or larger environmental hazards.

The fact that packaged waste moves across the social system as it changes hands creates a need to control a wider spectrum of human behavior, a necessity made clear to the American public in recent years by the increased incidence of midnight dumping. The desire for increased control has given rise to "cradle-to-grave" tracking schemes in most hazardous waste regulatory systems. These commendable attempts to delineate a chain of responsibility from the generator to the transporter to the final site of disposition represent a step in the right direction. Present programs, however, give only an illusion of social control, since a generator wanting to circumvent the system often simply does not declare the waste at the outset. This lack of reporting remains a serious short-circuit in the system. Furthermore, in many cases the regulatory agency lacks the resources to fully check the thousands of forms generated each year, some of which simply pile up in back rooms at government offices.

This need to control, more importantly, implies more direct control of industrial processes themselves, since regulators need to know the composition of waste, how and where it is produced, where it is sent, and how it is managed, not just the emissions from final processing and dispersal. This increased need for intervention runs counter to strong traditions of private property in the United States, where the sanctity of an industrial process is not under the purview of government. That this emphasis on private property delays needed reforms can no longer be ignored. The next chapter reviews the rigor with which European nations aggressively encouraged waste reduction schemes to lessen the burdens inherent in the management of packaged wastes.

Finally, a packaged waste creates a new dilemma whenever it crosses national boundaries. Conventional cross-border pollution, such as acid rain, involves dispersal of pollutants over large areas by natural processes. Controls on the source of the pollutants in the state of origin can mitigate the effects in other jurisdictions. For off-site disposal of hazardous waste, however, the environmental impacts are primarily experienced at the point of ultimate disposal. Thus, if the jurisdiction where the waste is generated allows the waste to be transported across its borders, it loses control over the impacts of the eventual management of that waste. Cross-border transfers of hazardous waste are given special attention in Chapter 7, where both recent scandals and their resultant international agreements are assessed in the context of government's increased role in the waste sector.

Waste Management Is an Invisible Service

Europeans also recognized, from their earliest programs, that a major difference between the hazardous waste management industry and manufacturing industries is that the resource inputs in the waste industry are of negative value. While most manufacturing industries meticulously monitor their inputs (since they pay premium rates for these raw materials), the waste industry is paid specifically to take its inputs away. Moreover, it is usually paid before it converts these materials into managed forms (that is, treatment residues). Thus the waste industry receives both its money and its inputs at once, but has inherent incentives to be concerned only with the former.

This structural property affects the regulatory control of wastes in two ways. First, off-site waste management is a service that is invisible to its purchaser. The treatment, storage, or disposal facility operator has little inherent incentive to complete the service, since he receives both the wastes and the payment before performing the service. Moreover, since waste also lacks value from the generator's point of view, it is usually not subject to quality control. As a result, waste quality varies widely in chemical composition and physical form. Different types of waste are often mixed together by the generator or handler, making treatment far more difficult. This variability in the waste stream is a major problem for American operators of hazardous waste management facilities, particularly high-tech treatment facilities trying to compete with land disposal practices, since the latter often accept any type of mixture or waste variation.

The abuses caused by the invisible nature of the service, such as the many instances of midnight dumping, have led to a profound distrust of private waste handlers. In response to this structural predicament, some European countries have chosen government ownership of waste treatment facilities. By examining the prominent European cases of government ownership, Chapter 5 evaluates the strengths and promise of these developments for America. Furthermore, the prospects of translating these discoveries to American political culture are examined in our concluding remarks.

PROSPECTS FOR AMERICA: THE KEY QUESTIONS

This book does not assert that management of hazardous waste in Europe is perfect. European nations are not without their Love Canals and groundwater problems. The strict Dutch laws that prohibit dumping, for instance, stemmed from the discovery of over 1,000 old chemical dumpsites that may be dangerous to human health. In 1980 268 families had to move from a housing development built above the Lekkerkerk dump—a former marsh filled with rock, gravel, and drums of chemical wastes—while cleanup crews began the long, expensive task of decontaminating the area.

Some hazardous wastes still fall through the cracks in Europe, owing largely to inconsistencies in the regulation of hazardous wastes among European nations. Countries where standards are lax have become havens for wastes from nations with strict requirements, posing problems for governments investing in expensive treatment facilities. Great Britain and Belgium still allow hazardous wastes to be dumped in the ground, for example, while East Germany has created a huge dump near the West German border that accepts any waste from any country at cut-rate prices designed to lure hard Western currency.

The most obvious, but often overlooked, lesson from Europe is that detoxification technologies exist and reduction strategies are possible that virtually eliminate the need for land disposal of untreated chemical wastes. Safe management of hazardous waste in this country need not await the advent of new, "space-age" methods. The technology exists. What we need is a transition to fully committed government policies that facilitate the use of these technologies. This is not possible without first restructuring America's response so that it corresponds more directly to the intrinsic properties of waste management.

European innovations have raised some critical issues for American managers and policymakers. In the face of mounting liability and insurance rate explosions in the industry, have we decided, in effect, that completely private waste management creates too great a risk to the public and the environment? What might happen if the government both sites and finances a hazardous waste facility and then regulates its operating excellence? Would we have stronger or lesser leverage to upgrade the control of these significant risks?

Government does intervene in managing some waste in the United States, such as municipal garbage, but those practices are certainly not the norm. Sustained and blatant government intervention is more common in Europe—and more easily tolerated by European industries, large and small. This kind of government ownership may eventually be tolerated in the United States because federal and state governments have already assumed some of the immense expense of remediation of past mistakes. Furthermore, because European governmental organizations play a direct role in managing toxic waste facilities, industry must decipher and comply with a less complex maze of regulations. As a result, Denmark and two West German states for instance, have avoided America's two most paralyzing regulatory battles—deciding precisely what qualifies as a regulated hazardous waste, and determining what technologies should be employed to ensure public safety and environmental quality.

Who will pay for hazardous waste treatment facilities? Some European nations have heavily subsidized the construction and operation of treatment facilities to lower costs to industry. When only short-run costs are taken into account, the costs of treatment technologies are generally higher than the price of conventional land disposal. Governments in this country can aid the shift to treatment facilities by financing their construction or providing low-interest loans or tax breaks to private waste management firms. We must be careful, however, not to subsidize waste management so much that we discourage innovative efforts to reduce the generation of wastes at the source. A number of chapters address this delicate question of balance (see 2, 3, 5, and 6).

How can we site treatment facilities in the United States? Most of the European facilities operating today were sited before risk became catastrophe—that is, before land disposal of hazardous wastes created the crisis that today is recognized the world over. As a result, public acceptance was quicker, more efficient, and less costly to European businesses. This latter question, which is examined at length in Chapter 8 and Appendix 4, is especially pressing, because even the latest high-tech European treatment facilities have not been immune to public opposition.

As Americans acknowledge that there are practical alternatives to land disposal, a further series of questions will arise. What degree of government leadership is required to shift America to strategies based on treatment and recycling (see Chapters 4, 5, and 6)? How can we encourage waste reduction to the extent found in European industries (see Chapters 2 and 4)? Is it feasible to build an infrastructure that fosters public-private joint ventures? What can we do to better address the broader challenge of toxic chemical risks? We offer *America's Future in Toxic Waste Management* to help frame the debate about each of these issues. Moreover, we see each of these issues presently emerging as the central environmental questions facing most industrialized nations until the end of this century.

The lessons derived from Europe in this book are merely transitional devices. The search for more pervasive solutions must continue. Waste reduction, recycling, and treatment are fundamental ideas whose time has come. Yet at a deeper level they are also transitional. Ultimately Americans must revise their view of nature itself. America's 200–year-old myth of the unending resilience of its soils will die only with great difficulty, as this last decade of trauma over toxic wastes forcefully attests. Gaining a more robust understanding of nature's limited capacity to assimilate waste takes time and a new view of humanity's place in the global commons.

NOTES

1. U.S. General Accounting Office, *Illegal Disposal of Hazardous Waste: Difficult to Detect or Deter,* GAO/RCED-85-2, February 22, 1985, pp. i-iii.

2. Readers particularly interested in these problems inherent in hazardous waste rule-making should see Bruce Piasecki's earlier book, *Beyond Dumping: New Strategies for Controlling Toxic Contamination* (Westport, Conn.: Greenwood Press, 1984).

Part 1

Minimizing Risks through Technology

2

Waste Reduction Strategies: European Practice and American Prospects

GARY DAVIS, DONALD HUISINGH, AND
BRUCE PIASECKI

THE PROMISE OF POLLUTION PREVENTION

Since the dawn of industrialism nations have struggled with the threat of industrial pollution. In the late 1960s and early 1970s a number of nations that recognized the extent of the threat embarked upon ambitious programs for pollution control. These programs focused on cleaning up what was coming out of the stacks or pipes of industrial facilities through the setting of emission and effluent limits that required the use of available treatment technologies.

This twenty-year emphasis on controls at the end of the manufacturing process has resulted in disappointingly slow and costly progress. Most industrial firms have followed government down this end-of-the-pipe path by installing expensive treatment units to meet the regulatory standards. Unfortunately, most of these treatment units, when they worked properly, either transferred the problem from the water to the air or from the air to water or resulted in the production of hazardous residues that were then disposed of on the land. In the current decade the shift of pollutants to the land weighs heavily on society. Many of the early standards resulted in the reduction of conspicuous pollutants such as the discharge of biodegradable organics into rivers that were depleting oxygen levels or the emission of particulates to the air, but ignored more difficult and more serious problems like discharges of toxic organic chemicals and heavy metals. Moreover, consistent and reliable enforcement of these controls has also proven to be elusive because of lack of government resources, heavy paperwork burdens, and the existence of too many opportunities to evade compliance in our self-monitored systems.

Perhaps the most discouraging result of this end-of-the-pipe approach has been the fostering of the idea that economics and the environment are in a direct and vicious conflict. Although the pollution control approach has produced measurable environmental benefits, the overall economic inefficiency of end-of-the-pipe solutions is becoming ever more apparent.

Herein rests the promise of waste reduction. Many industrial firms now find—in the face of mounting environmental liabilities—that when they reassess waste as indicative of inefficiencies in their processes, a new approach based on pollution prevention is called for, whereby the economics of production can be improved at the same time the environment is protected. The 3M Company coined the slogan "pollution prevention pays" to proclaim this new approach. This chapter examines the promise of this paradigm shift by exploring the role of government in stimulating all industries to incorporate the economic and environmental benefits of waste reduction into their facility designs and corporate decision making.

Several European governments grasped this basic discovery in the 1970s and began taking a leadership role in making the shift possible. By using a range of policy measures, they fought hard for fundamental, not incremental, changes in industrial design. While the United States has recently begun to follow this lead by initiating state and national policies on waste minimization, American business remains seriously segmented in its response to waste reduction. What some view as sensible many still view as inconvenient compared to the continued ease of their present processes and disposal options.

The strength of the example of European programs rests in the extensive institutionalization of the concept. It is impossible, however, to determine whether European nations have in fact reduced pollution per unit of industrial output to levels significantly below those of the United States, since adequate or reliable data to provide such a comparison simply does not exist. Nonetheless, what remains vividly clear in Europe is the exceptional level of effort and financing several nations have provided for these reforms, given their size and economies. The telling lesson from these nations is that sustained government involvement is a necessary ingredient in the shift to pollution prevention strategies, just as it has been the ingredient that has perpetuated end-of-the-pipe pollution control strategies.

The European lead in pollution prevention is built upon four foundation stones: (1) the scarcity of vacant land away from centers of population for land disposal of hazardous wastes; (2) an intense concern about groundwater contamination, due to the high level of dependence on groundwater for drinking water; (3) the post–World War II material scarcity, which created a conservation ethic, giving rise to societies where recycling is an ingrained habit; and (4) the traditional high level of government/industry cooperation in all areas of industrial policy and technology development.

The first two factors need little elaboration, except to point out that the population density in Western Europe is many times greater than in most American states and that in West Germany, for instance, over 70 percent of the people receive drinking water from groundwater. The third and fourth factors are deeply cultural and institutional.

The conservation ethic in Europe is most evident in the recycling of consumer goods and packaging materials, such as returnable bottles, newspaper recycling, and used motor oil. This translates into the existence of government programs for recycling of materials, despite the fact that recycling may not be, in the terms of short-term economics, competitive with disposable containers or virgin products. The European emphasis, in contrast, is on longer-term sustainable economics.

The cultural roots of government/industry cooperation in Europe run deep, from the ancient authoritarian rule of monarchies to the modern socialistic nationalization of industries. The institutional manifestation of this in the field of hazardous waste management is the significant level of government support for pollution prevention that several countries have instituted as a key part of their general industrial policy. Pollution prevention or "clean technology" programs are seen not only as environmental improvements, but also as an important means of encouraging industrial innovation. This fundamental reemphasis of waste reduction as a force for revitalization of industry makes remarkable sense. These two general institutional lessons, the emphasis on longer-term economics and the willingness to encourage pollution prevention as a path for further industrial innovation, create the backdrop to the detailed examples of policies and programs next discussed. While we cannot duplicate the cultural foundations of European pollution prevention programs, we may at least be able to select institutional lessons for application in the United States.

Whether it is called pollution prevention, waste reduction, waste minimization, low- and non-waste technology, or clean technology, the essence of the concept is the same. Pollution, or waste, is eliminated or reduced at the source within industrial processes rather than "controlled" at the end of the pipe or stack. The terms "low- and non-waste technology" and "clean technology" are the most widely used in Europe and are broader than "waste minimization" or "waste reduction" as presently used in U.S. programs.

The United Nations Economic Commission for Europe (ECE) defines "low- and non-waste technologies" very broadly as "the practical application of knowledge, methods and means, so as—within the needs of man—to provide the most rational use of natural resources and energy, and to protect the environment."[1] The ECE Declaration on Low- and Non-Waste Technology and Reutilization and Recycling of Wastes is a strong policy statement that encompasses reduction of pollution at the source, on-site recycling, off-site recycling, more efficient use of raw materials, and energy conservation.[2]

The Commission of the European Communities succinctly defines "clean technology" as "a technique to produce a product with the most rational use of raw materials and energy, at the same time reducing the amount of polluting effluents in the environment and the quantities of wastes produced during manufacturing as well as during the use of the manufactured product."[3]

The force of these broad definitions rests in their advocacy of a design triangle, whereby the desires to conserve energy, save materials, and reduce waste join in a shared dynamic. "Clean technology," then, means consciously making decisions about the products a nation produces and the manner in which they are produced. This design decision, moreover, is based upon a whole array of primarily economic motives mixed with a set of broad-based environmental factors. By defining the goal as clean technology, the Europeans are truly starting at the front end, not at the tail end of the process, which America's focus on waste reduction still tends to encourage.

Among the benefits from the pollution prevention approach are the following:

1. Reduced worker exposure to hazardous chemicals
2. Less hazardous waste transported on the highways
3. Reduced need for off-site hazardous waste facilities with their attendant environmental and political problems
4. Increased industrial efficiency through raw materials and, in some cases, net energy savings
5. Reduced liability for leaking dumpsites and inadequate off-site disposal practices
6. Avoidance of increasing disposal costs and administrative costs inherent in cradle-to-grave management

Despite the other staff and production demands competing for resources within a corporation, the benefits of pollution prevention are gaining in value. If defined as a tool for improving the overall bottom line rather than merely as another technical fix to improve plant yields, waste reduction decisions can move higher up the corporate ladder into the financial realms of upper management, generating enthusiasm among a company's highest fiscal and risk-control managers.

THE PRACTICE OF POLLUTION PREVENTION

There are four basic technical approaches to reducing hazardous waste generation:

1. Source segregation or separation
2. Process modification
3. End-product substitution
4. Recycling and reuse

The first three approaches occur within the generator's plant, while the fourth can also occur in off-site commercial facilities or between industrial firms. No single approach or specific technique can be said to be the best one, since any or all of the four approaches may be applicable for a given generator's waste stream.

Source Segregation or Separation

Source segregation is the simplest and usually least costly waste reduction method. Hazardous waste generators who now use land disposal may only have to reorient their thinking to take advantage of simple source segregation techniques. It can involve a change in operating procedures as simple as segregating two waste streams that can each be recovered when separate, but are difficult to recover when mixed. For example, electronics firms often generate both waste halogenated solvent and non-halogenated solvent wastes. When these solvents are separate they are easy to recover by distillation processes, but when they are mixed, distillation becomes much more difficult. Source segregation can also be simple attention to housekeeping, such as maintaining process equipment to prevent leakage of hazardous chemicals.

Process Modifications

Process modifications involve changes in design and operation of plant processes. These include changes in raw materials used and in reaction conditions and procedures as well as the retrofitting or replacement of process equipment. Process modifications involve an intimate knowledge of the process and of process technology in general. In-house engineers, consultants, and equipment suppliers are usually all involved in planning and implementing the changes.

In a typical manufacturing plant, process modifications are usually considered for the purpose of increasing profits rather than for the prevention of waste generation. As waste disposal costs increase and raw material costs rise, process modifications are being made more and more for the dual purpose of reducing waste generation and increasing profits.

End-Product Substitution

End-product substitution for hazardous waste reduction is the replacement of a waste-intensive product with one for which production or use does not involve the generation of as much hazardous waste. Examples include the replacement of PCBs as electrical insulating fluids with silicone oils and the replacement of cadmium electroplating with zinc. Since waste-intensive products are often more hazardous in and of themselves, product substitutions can have the additional benefits of protecting workers who manufacture them, the public who uses them, and the environment upon which all life depends.

Recycling and Reuse

Recycling and reuse is the practice of recovering usable materials from wastes. The simplest application is reusing a waste from a process directly as a raw material in that or another process. When this is done between firms, it is called waste exchange.

Most wastes require some type of processing before they may be reused. The most common example is the recycling of waste organic solvents by distillation to separate clean solvent from impurities. Other commercial recycling techniques include carbon adsorption and solvent extraction to recover organics from water and ion exchange and reverse osmosis to recover metals from electroplating rinse waters.

Recycling and reuse are performed both on-site and at off-site commercial facilities that handle similar wastes from a number of generators. Commercial facilities that recycle organic solvents are the most common, but recycling facilities for other wastes, such as waste acids and heavy metals, exist as well. Waste exchanges are operated either as clearinghouses for information concerning waste materials available for reuse or as brokerships, where an intermediary takes possession of the wastes and transfers them to a purchaser.

Hundreds of industrial firms in Europe and the United States have implemented the types of measures listed above to reduce waste generation. Some of the outstanding examples from Europe include the following:

Leather-Tanning In-Plant Recycling Measures

A leather-tanning firm in France has implemented a series of measures for in-plant recycling of effluents. These include reduction of the organic pollution from the preliminary stage of preparing the hides; recycling of the chromium-containing effluents from the tanning process; and recycling of the effluents from the baths used for tanning. These recycling measures have yielded substantial reduction in discharges of organic and chromium-containing effluents. In addition, chromium recycling generates significantly less chromium-containing sludges than the conventional treatment method, minimizing waste disposal problems. These recycling measures also yield savings in raw materials that help to offset the costs of operating the recycling systems.[4]

Low-Emission Paint-Drying Technique

An innovative thermoreactor paint-drying technique has been developed by a French company (SUNKISS). This technique has been installed at metal-finishing operations for products ranging from small metal objects to cars and locomotives. The use of this technique by a French company (Alstholm Atlantique) on two of its metal-painting lines has yielded the following environmental and economic benefits:[5]

1. A 99 percent reduction in emission of evaporated solvents that are destroyed in the catalytic heating/drying process involved in the thermoreactors

2. Elimination of the explosion risks usually associated with drying operations in the presence of solvent fumes

3. A 99 percent reduction in drying time, which increases throughput and facilitates automation on the painting lines

4. An 80 percent savings in the energy requirements for the drying operations, which yields annual savings of 1.1 million French francs. This resulted in a pay-back period of only two months on the cost of purchasing and installing the thermoreactors.

Dornier Ion-Exchange Columns

The Dornier ion-exchange columns are a simple but effective method of reducing the discharges of toxic metals from electroplating operations. Metals in effluents are adsorbed onto ion-exchange resins, which are collected and regenerated at central facilities. The compactness and flexibility of the process makes it particularly suitable for small and medium-sized firms. The use of the ion-exchange columns yields improvements in product quality, savings in plating materials, and lower water usage. Differing water charges and effluent charges yield payback periods that vary depending upon the specific circumstances of the individual metal plater. Nonetheless, these Dornier ion-exchange columns have been extensively applied in the electroplating industry, especially in Germany and France.[6]

Recovery of Hydrochloric Acid from Acid Pickling of Steel

The use of hydrochloric acid instead of sulfuric acid for pickling steel allows for a closed-loop process in a system developed in Austria. The iron chloride produced during pickling is decomposed by heating in a furnace to iron oxide and HCl gas, the acid gas is dissolved in water for reuse, and iron oxide is recycled to the blast furnace. The process allows for nearly a 90 percent reduction in acid usage and eliminates sludge generation from neutralization of the acid as a treatment process. These savings make the capital investment profitable.[7]

Recycling Etchant in Chrome Plating of Plastic

Chrome plating of plastic requires preliminary etching that is performed by dipping the plastic parts in a sulfuric/chromic acid bath. Prior to the institution of the recycling process by this French firm, the whole bath was sent to a hazardous waste treatment facility when the hexavalent chromium content of the bath was depleted. The recycling process utilizes electrolysis to reoxidize trivalent chromium in the bath to hexavalent chromium, permitting recycling of much of the bath. The firm now sends only four tons per year of spent etching bath off-site for treatment instead of the fifteen tons per year previously shipped off-site, and due to savings in sulfuric and chromic acid losses and avoided treatment costs the capital investment paid for itself in fifteen months.[8]

EUROPEAN POLICIES TO ENCOURAGE WASTE REDUCTION

While waste reduction has occurred and would continue to occur in selective segments of industry without government initiative, government leadership can greatly facilitate this shift in emphasis. The generation of waste is the result of several choices made by industrial firms in the production of goods and services, such as the choice of product to produce, the raw materials and processes used, and the degree to which recycling is practiced. Although these choices are often based upon simple economics, they are also heavily influenced by noneconomic factors, such as access to information, availability of technology, institutional inertia, liability considerations, capital resources, and government regulations. Government initiatives can not only shift the economics, but can also exert influence through each of these non-economic factors.

There are five basic types of government initiatives in pollution prevention that are being used in European countries:

1. Education and information transfer
2. Research, development, and demonstration
3. Direct technical assistance
4. Economic incentives
5. Mandatory measures

The European governments surveyed are using a mixture of these initiatives to encourage waste reduction, some in comprehensive programs with a strong institutional focus.

The Power of Properly Placed Information

Education about the promise of pollution prevention, as well as reports on specific measures for waste reduction, has been one of the most consistent roles of government in encouraging waste reduction. The activities of most international agencies, for instance, have centered primarily on education and information transfer.

The United Nations Economic Commission for Europe has been the most involved in promoting the concept internationally. The ECE sponsored the first International Conference on Non-Waste Technology held in Paris in 1976. In 1979 the ECE adopted a detailed Declaration on Low- and Non-Waste Technology and Reutilization and Recycling of Wastes.[9] In the four-step declaration, the ECE

1. declared its intent to protect man and the environment and to use resources rationally by promoting low- and non-waste technology and reutilization and recycling of wastes;

2. recommended national actions to (a) promote research and development activities on low- and non-waste technologies; (b) provide incentives for use of low- and non-waste technologies by studying and implementing economic incentives, regulations, and standards to overcome constraints to greater utilization; (c) encourage transfer of low- and non-waste technologies between industries and industrial sectors; and (d) include the concept of low- and non-waste technology in educational programs at all levels;

3. recommended international activities to (a) support research and development through cooperative pilot projects and through development of a unified classification of waste streams; (b) exchange scientific and technical information by compilation of a compendium of low- and non-waste technologies and by conducting international seminars; (c) organize activities concerning international waste exchanges; and (d) organize international postgraduate courses on low- and non-waste technology;

4. recommended that a body on low- and non-waste technology be set up within the framework of the ECE.

Since that time, ECE and its member nations have been working to carry out these recommendations. Another international symposium was held in Tashkent, USSR, in 1984. The ECE published a six-volume compendium on low- and non-waste technology beginning in 1981, which listed over 80 examples of successful pollution prevention efforts by European industrial firms[10] and also published a compendium of lectures by experts in low- and non-waste technology in 1983.[11]

The European Community (EC) has also been considering the promotion of clean technologies in its activities. The European Commission, the administrative arm of the EC, has performed a series of studies on industrial sectors to determine the technologies in use and the potential for use of clean technologies. The commission also organized a European Seminar on Clean Technologies that was held in The Hague in 1980.[12] Activities to support information transfer on the international level are continuing. There is currently a proposal in the ECE to set up an international center for low- and non-waste technologies.[13]

While the international agencies popularized the validity of the concept of clean technologies, most of the European nations surveyed set up distinct government offices for the specific implementation of clean technologies. Most of these institutions transfer information about waste reduction directly to industry in some fashion. To enable them to reach a wide range of users, many of the institutional settings for information transfer activities are deliberately separate from the environmental regulatory functions of government.

The Clean Technologies Office of the Danish National Agency for Environmental Protection actively disseminates information on clean technologies. For each research and development project supported by the office, detailed reports are made available to the public, and the office publishes a general report of results each year. At times, this office has used the results of its industrial sector research and the threat of more stringent regulations to put pressure on firms to implement clean technologies in order to go beyond the requirements of existing environmental regulations.[14]

The French pursue waste reduction from a number of institutional fronts. Three principal government programs promote waste minimization and clean technologies. The two national agencies are the Mission for Clean Technologies (Mission Technologies Propres), in the Ministry of the Environment, and the National Agency for the Recovery and Elimination of Waste (Agence Nationale pour la Récupération et l'Élimination des Déchets). On the regional level, six river basin financial agencies (Agences Financières de Basin) are involved in the encouragement of reductions in water pollution and hazardous waste generation. None of these agencies or programs has direct regulatory authority.

The Mission for Clean Technologies has been extensively involved in education and information transfer concerning clean technologies. In 1982 the mission published a compendium of clean technologies with over 100 examples from all industrial sectors. This compendium has been translated into English and published by Dr. Michael Overcash at the University of North Carolina.[15]

In 1984 the river basin financial agencies and the regional offices of the Ministry of Industry and Research implemented a national database of clean technologies for each industrial branch, which reportedly includes performance information and cost evaluations. All industrial firms have access to this information through these regional offices.[16] In addition, the French Ministry of the Environment awards a yearly prize to outstanding industry projects implementing clean technologies.[17]

The Department of Clean Technology in the Dutch Ministry of Housing, Physical Planning, and Environment has plans for an even more highly visible information institution. The agency will transfer research findings and other information on clean technologies through the establishment of a national information center set up jointly by industry, the national government, and the Dutch Chamber of Commerce and Industry. Efforts to transfer information on research and development projects are currently made through trade organizations representing various industrial sectors.[18]

Research, Development, and Demonstration: Overcoming Technical Barriers

In an effort to overcome technical barriers to waste reduction, the European Community is the only international organization to provide direct economic incentives for clean technologies. In 1984 the European Council of Ministers initiated an Environmental Fund to be allocated to projects that demonstrate broadly applicable waste reduction innovations in industries that generate large amounts of pollution or discharge wastes with significant environmental hazards. The projects funded in 1985 were selected by a committee of experts representing the ten members of the EC. The EC hopes to increase the allocation of resources for this purpose during the 1986–1987 fiscal year.[19]

Specific national initiatives show the impressive level of support. The Austrian government, for instance, established an Environmental Fund in 1983 to foster research and development in non-waste technologies. This fund is administered under the Austrian Federal Economic Chamber. A multidisciplinary team decides which applicants receive funding from the Environmental Fund.

The fund has been used for add-on pollution control equipment, but more recently the focus has been on clean technologies. A grant of 20–30 percent can be provided for demonstration projects, or a low-interest loan can be provided for capital costs of new equipment. The fund, which was created from general tax revenues, totaled 1 billion Austrian schillings (about $70 million) in 1985. Approximately 12 million Austrian schillings (about $850,000) were invested in the period April 1984 to May 1985.[20] A recent document describes 23 newly implemented clean technologies, as well as two projects under development, which have been funded by the Environmental Fund. Examples include a closed-loop hydrochloric acid steel-pickling process and production of aluminum fluoride from waste gases from phosphoric acid plants.[21]

In Denmark a Clean Technology Fund was established by direct amendment to the Act on Recycling, Reuse, and Reduction of Waste. This 1984 program stimulated significant activity due to extensive publicity and a growing recognition of the benefits of clean technologies by Danish industries. The stated purpose of the new law echoes the concept of a design triangle. The fund exists

to prevent and combat pollution by providing a basis for reducing the impact on the environment caused by waste and to secure a socially appropriate use of raw materials, products, materials and residuals by recycling or by using cleaner technologies creating less pollution than the technology hitherto applied.[22]

Under a multiple funding approach, aid can be granted to research and development projects for (1) process changes for waste reduction; (2) changes in the use of raw materials to reduce waste generation; and (3) changes in product design or product substitution.

In recognition of the societal benefits of clean technologies, R&D projects that benefit industry in general or society as a whole may receive grants of up to 100 percent of the costs. For R&D projects that benefit only one firm, up to 75 percent of the costs may be covered by a grant. These grants can include the development of more efficient processes, innovative production equipment, and new products that are easier to recycle.

In order to receive the aid, applicants must demonstrate that the investment will create a permanent increase in the recycling of products, materials, or residuals, thereby significantly reducing their impact on the environment. Decisions on eligibility for aid are made by a ten-member council appointed by the Minister for the Environment, which includes representatives from industry, labor, local governments, and other government agencies.[23] This mix ensures that a diverse array of projects will be funded and that, down the road, there will be a greater degree of participation and dissemination of information.

For the first year, approximately $3 million was distributed to projects for reducing all types of pollution and waste. So far industry has responded favorably.[24] Hazardous waste projects funded during the first year include development of an information transfer system for recycling and clean technologies for the electroplating industry, and the promotion of industrial waste recycling by a local authority. The funding was allocated to private industry, technical institutes, consulting firms, and local governments.[25] This willingness to fund a great range of institutions also enhances communication among the different users of the results.

In addition to the external research funded by the government, the Clean Technologies Office has performed industry-specific studies on clean technologies. From 1983 to 1985 studies have been performed for the steel industry and for the metal-finishing industry. The reports not only include detailed descriptions of pollution problems and of the cleaner technologies used to solve the problems, but also carefully explore the economics of these cleaner technologies. Projects are under development on ways to implement cleaner technologies on a widespread basis and to establish information systems that can fit more effectively into international systems.[26]

The French have several programs for the financial support of research and development in waste reduction. In 1979 the French Mission for Clean Technologies was established in the Ministry of the Environment to promote pollution prevention by focusing the non-regulatory activities of the ministry in this direction. The mission provides financial aid to demonstration projects for clean technologies that reduce pollution (air, water, solid and hazardous waste) through product substitution, process changes, and internal recycling. The assistance provided is in the form of grants of up to 10 percent of project costs for innovative approaches that are transferable throughout the specific industrial sector.

Applications for assistance are reviewed by an interministerial group for the coordination of actions related to clean technologies. From 1979 to 1982 the mission gave over 19 million francs (about $3 million) of subsidies to 54 projects representing 524 million francs (about $82 million) in investments in clean technologies.[27]

In 1975 the French National Agency for the Recovery and Elimination of Waste (ANRED) was established by law as a "state-owned public service of an industrial and commercial nature responsible for facilitating waste disposal and recovery operations, or for carrying out such operations in the public interest when private or public means are insufficient." Thus the agency was set up totally as a public service, without any regulatory authority. One of the stated goals of ANRED is to "promote information campaigns to make the general public and industry aware of recycling and proper waste management."[28] As a result of this mandate, ANRED aggressively promotes waste reduction and re-cycling through films, television and print ads, and literature.

In the area of economic incentives ANRED operates by giving grants or low-interest loans to projects for waste minimization, recycling, and waste management in the field of solid and hazardous wastes. Projects supported include those for research and development as well as those for full-scale facilities.

In 1983 ANRED allocated over 1 million francs (about $150,000) in grants and loans for research projects for hazardous waste reduction and on-site recycling of hazardous waste. Demonstration projects for hazardous waste reduction and recycling totalled 5.2 million francs (about $819,000). ANRED has funded projects for recovery of mercury, for recovery of lead and zinc from metallurgical wastes, for regeneration of solvents, for recycling of methanol and catalysts by a chemical firm, for reuse of sludges from paper mills, and for regeneration of used oil.[29]

Other national agencies can also provide research and development assistance for waste minimization and clean technologies. The Research Service (Service de la Recherche) within the Ministry of the Environment provides grants of up to 50 percent for research on clean technologies applicable to whole industry sectors. Their emphasis is on process modification and product substitution. Between 1979 and 1982 this service awarded 17 million francs (about $2.7 million) of grants to 67 projects representing 41 million francs of total research costs.[30]

The National Agency for Encouragement of Research (Agence National pour la Valorisation de la Recherche [ANVAR]), part of the Ministry of Industry and Research, provides assistance to demonstration projects to promote technological progress. The assistance is in the form of a loan that must be reimbursed only if the project is successful. The loan can be up to 50 percent of the project costs. In 1982 10 percent of the assistance for innovation was given to projects for the reduction of pollution.[31]

In 1979 the Dutch government established a special Committee on Environment and Industry that is responsible for fostering joint efforts among governmental agencies, industry, and academia with the explicit goal of promoting more research on clean technologies. The five Dutch ministries involved in this effort have established joint funds for subsidizing research and development activities on clean technologies.[32]

To date, this Dutch Committee on Environment and Industry has facilitated approximately 200 research, development, or demonstration projects for clean technologies. Approximately 20 million Dutch guilders (about $8 million) per year is allocated for these efforts. Given the size of the Netherlands, this amount is very impressive. Funding is provided to industry, research institutes, and governmental agencies.

Some examples of projects funded include replacement of cadmium plating with aluminum, the production of low-cadmium phosphate fertilizers, and the use of biotechnology to reduce waste generation. A special research program on membrane technologies for recycling of pollutants was also recently established.[33]

The West German Federal Environmental Agency (Umweltbundesamt [UBA]), established in 1974, funds research and development projects in all fields of pollution control, including waste reduction, recycling, and waste management. The Section for Waste Minimization in the UBA can pay up to 50 percent of private research for new processes to reduce waste generation. The section prioritizes the research that it sponsors based upon the need to develop alternatives to the current management methods for particular waste streams. For instance, recent research has been directed toward waste minimization and off-site recycling measures for those wastes that have been either dumped or incinerated at sea, since the Germans have decided to eliminate ocean disposal by 1990 (see Chapter 4 for more details). Often, despite the fact that the UBA has no regulatory authority, the products of successful research are used as a tool of persuasion to change the practices of a whole industry.[34]

The UBA also performs in-house research or contracts for waste minimization research projects of a generic nature. Such research is focused on the waste streams of the greatest quantity and toxicity, often those most in the public eye. Recent sectoral studies include a systematic screening of wastes generated in and waste minimization measures for the chlorinated organic chemical industry. A high-visibility study of the occurrence and prevention of dioxin-contaminated wastes has also been completed by UBA.[35] In 1983 the UBA provided a total of DM 19 million (about $8.5 million) to assist research in the field of hazardous waste minimization and recycling.[36]

In Sweden grants of up to 50 percent of capital costs may be given to demonstration projects, which increasingly include pollution reduction measures of all types. In 1983–1984 the Swedish National Environmental Protection Board distributed about $10 million in support for new environmental technologies.[37] No breakdown is available on the percentage given for waste minimization.

The Swedish government has also funded a major independent multidisciplinary research organization called TEM, located at the University of Lund. The organization's prime responsibilities are to

1. develop new low- and non-waste technologies

2. develop new and more efficient methods of recycling previously used resources;

3. develop a comprehensive understanding of the social and economic factors involved in low- and non-waste technologies in comparison with previously used technologies.[38]

This concept of capitalizing on the expertise of universities deserves serious exploration in the United States.

Direct Technical Assistance: Government as Consultant

Only a few European governments provide direct technical assistance to waste generators to help them reduce waste generation; most leave this kind of work to the professional engineers and consultants. The French ANRED has a staff of engineers who provide assistance to industry and municipalities on measures to reduce waste generation. ANRED's technical staff budget for 1983 was about 7 million francs ($2.6 million).[39] The river basin agencies and the regional offices of the Ministry of Industry and Research also provide free direct technical assistance. Information was unavailable concerning whether this technical assistance is provided through on-site waste audits.[40]

Although there is no organized program for direct in-plant technical assistance for waste minimization in West Germany, the members of the Federal Environmental Agency (UBA) staff often work closely with their counterparts in industry to solve waste generation problems through the research that UBA funds. This informal cooperation between UBA and industry is facilitated by the fact that UBA is not a regulatory agency.[41]

Economic Measures: Creating the Market Signals

In addition to funding research and development projects for waste reduction, several European nations provide other economic incentives and disincentives, such as grants and low-interest loans for capital expenditures, tax breaks for capital investment in waste reduction, and taxes on waste generation.

Since 1975 the Danish have made use of subsidies for environmental protection, allocating over 250 million Danish kroner (about $34 million) in the ten-year period between 1975 and 1985, mostly for pollution control technologies. In 1980 the Act on Subsidies for Environmental Investments was amended by the Parliament to shift the emphasis to clean technologies. The idea was to determine whether companies were interested in making radical and often expensive changes in production technology to meet or exceed environmental standards rather than continuing to make use of end-of-the-pipe solutions. This legislation provided that a new plant or modification to existing processes that was designed to reduce pollution (air, water, and land) at the source could receive up to a 25 percent subsidy from the national fund. Decisions on eligible projects were made by a board including representatives of the Ministry of the Environment, municipalities, and industrial associations.[42]

In addition to this subsidy program in Denmark, the 1984 amendments to the Act on Recycling, Reuse, and Reduction of Waste described in the section "Research, Development, and Demonstration" created a Clean Technology Fund that can be used to subsidize capital investments in recycling. Grants up to 25 percent may be awarded for investments in recycling equipment and grants up to 75 percent may be awarded for waste collection equipment. It is telling that the Danish also include collection technology as part of the picture, since collection systems are often the weakest link in recycling programs.

The French ANRED provides assistance with the capital costs of full-scale facilities for recycling and waste management. These facilities are typically off-site commercial facilities. ANRED uses several different types of financial assistance. Where recycling or waste reduction projects are publicly owned or for the benefit of the public at large, ANRED may provide grants of up to 100 percent of the project costs. Where projects are of a commercial recycling nature, ANRED may participate in the technical risks of the projects. For instance, for recycling ventures where the market for recycled material is variable, ANRED invests in the venture, sharing the risks run by the firm and offering a partial guarantee against failure without becoming involved in the management responsibilities.[43] Such sustained government involvement in the recycling market has proven to be very effective. (See Chapter 6 on the French waste oil program.) Also on the national level in France, the Investment Fund for Modernization, administered by the Ministry of Industry and Research, gives low-interest loans of up to 70 percent of capital investments to increase productivity in industry. These have included important projects that reduce waste generation.[44]

The French river basin financial agencies use a combination of economic disincentives and economic incentives for waste reduction. The six public agencies tax water pollution in order to subsidize process modifications and treatment facilities to minimize or treat waste that might pollute groundwater or surface waters. Fees are collected from municipalities and industries discharging pollutants into rivers and streams as well as from water users. In 1978 alone effluent charges for the whole country totalled about $250 million.[45] The basin agencies provide a 10–20 percent grant or a low-interest loan to firms making process modifications that minimize water pollution and waste generation at the source.[46]

In addition to these subsidies, the French tax system allows for rapid depreciation of investments that prevent pollution, including process modifications and internal recycling. Fifty percent of the capital costs of a waste reduction project may be written off during the first year of operation.[47] Such an investment option in the United States would likely stimulate design reforms.

The Norwegian Pollution Control Authority has a grant and loan program that was begun in 1974 to encourage existing industries both to use cleaner technologies and to install pollution control equipment. In its first decade of operation, the program provided about $500 million in subsidies, a massive amount for the size of Norway's economy. There is now a bias in favor of projects that reduce pollution through process changes. Grants of 40–50 percent of capital costs and low-interest loans of up to 80 percent of capital costs are still available. Examples of projects funded include a process change for a fertilizer plant and a recycling process for sulfur liquors in a paper mill.[48]

The Swedish government also has used subsidies as a tool for environmental protection for many years. When the Environment Protection Act was passed in 1969, the government provided for grants of up to 25 percent of capital investments in pollution control equipment. During periods of relatively high unemployment in the 1970's the Swedes used environmental subsidies as a way to create jobs and provided subsidies of up to 75 percent for pollution control equipment. By doing this, they built a bridge between environmental reform and economic recovery. Between 1969 and 1977 $125 million in subsidies was provided to Swedish industry under this program.[49] In 1977 the general subsidies were eliminated, and since that time only demonstration projects for new and untried technologies can receive subsidies.[50]

The Netherlands is one of the few European countries to impose a tax on the generation of hazardous waste. The government uses this disincentive to waste generation, as in the state of California, to generate revenues for its waste regulatory agency. The tax is authorized by the Chemical Waste Act of 1976 and has been imposed upon generators of chemical waste who manufacture certain products or use certain chemical processes specified in the regulations. A tax is also imposed upon treatment, storage, and disposal facility licensees.[51] The Dutch are planning to eliminate this tax, however, because their information base on waste generation has proven inadequate.[52]

Mandatory Measures: Saying No to Waste

Some European countries have decided that mandatory measures are needed to reduce waste generation, although most of the mandatory provisions that have been enacted have not been widely enforced. It may be that the threat of their enforcement, alone, is enough to ensure the success of other government initiatives used.

The Danish 1984 Act on Recycling and Reduction of Waste gives the minister of the environment power to restrict the use of certain raw materials, additives, or intermediate products that prevent the recycling of paper and packaging materials. Although this provision does not apply directly to hazardous waste, it may eventually have a strong influence on the generation of hazardous wastes by encouraging product substitution. No restrictions have been imposed as of the end of 1985.[53]

In addition to this limited authority, the minister of the environment, pursuant to the Environmental Protection Act, may require that certain types of heavily polluting industries must receive prior approval before constructing a new plant or substantially modifying an existing one. The list of types of industries is fairly broad, containing 96 specific entries, ranging from steel mills, oil refineries, and chemical plants to slaughterhouses and crematoria. Applications under this provision must contain information about measures to be adopted to reduce pollution, and the regional councils have conditioned approval on the adoption of pollution prevention or pollution control measures.[54]

Finally, the Danish government has enacted a product control law, the Act on Chemical Substances and Products of 1980, which gives the ministry the authority to restrict the use of certain products that are dangerous to human health or the environment, and also gives the authority to require a deposit to be charged on certain products to encourage their return to the manufacturer or seller rather than disposal.[55] Under the authority of this act, the use of PCBs has been prohibited and uses of cadmium, asbestos, and chlorofluorocarbons have been restricted. The deposit authority has not been used yet, but the ministry is considering requiring deposits on batteries, paints, and tires in order to decrease hazardous air emissions from combustion of household waste in municipal incinerators.[56] Should this happen, the program would be of immense interest to local governments in the United States now planning mass-burn approaches to their solid waste problems.

The French Waste Law of 1975 established the general principle that management of waste must be performed in a manner to facilitate the recovery of usable material and resources. The law gives the Ministry of the Environment the authority to control the manufacture, importation, storage, or sale of products that generate waste. Regulations on production can be issued to facilitate the recovery of waste, or outright prohibitions of manufacture or use of certain products may be issued. The Ministry of the Environment established waste minimization as the first priority of its waste regulatory program in 1979, but available information shows that the ministry has primarily made use of economic incentives to encourage waste minimization rather than its considerable mandatory authority.[57] (For an exception to this, see Chapter 6 dealing with French waste oil programs.) This decision to encourage reform via fiscal stimulation rather than with the regulatory stick is an approach used throughout much of Europe.

The government of The Netherlands has available a formidable array of mandatory measures for waste reduction under its Chemical Waste Act of 1976, although, to date, these mandatory measures have not been put into effect. The chemical waste statute vests broad powers in the Minister of Housing, Physical Planning, and Environment. The section of the law on waste generation restrictions applies to chemicals that as waste are impossible or very difficult to manage as well as chemicals that as waste may be injurious to health or the environment if not properly treated, processed, or destroyed. Under the section the minister can adopt rules that contain an absolute prohibition to manufacture or market such goods or a requirement that such goods be manufactured or marketed in the special way set out in regulations. Section 33 of the act also grants the minister the power to order a generator of chemical waste to treat, process, or destroy its waste on-site.[58]

Under the Norwegian Pollution Control Act of 1981, as amended, the Pollution Control Authority may require the recycling of solid and hazardous waste by the generator.[59] In addition, pursuant to regulations promulgated under this law, hazardous waste generators must obtain permits from the authority. Conditions on recycling and reprocessing may be included in these permits.[60]

The Norwegian Product Control Law of 1977, which is similar to but more comprehensive than the Toxic Substances Control Act in the United States, has been used to restrict the usage of several substances that create severe environmental problems throughout their life cycles. In addition to banning PCBs, the Norwegian Pollution Control Authority has restricted the use of cadmium, phosphates in detergents, chlorofluorocarbons, and asbestos, forcing in each case the use of safer substitutes.[61] This willingness to use product controls as an encouragement for utilization of safer substitutes provides the surest route to pollution prevention, reducing toxic chemical exposure throughout the whole product life cycle from workplace to disposal.

The West German government has recently adopted changes to the Waste Law of 1972 that will make waste minimization and recycling an integral part of federal waste requirements and provide the German states with broad powers. The new Waste Law requires that when technically and economically feasible the generation of waste should be avoided and low-waste or non-waste processes should be used. When new production facilities or modifications to existing facilities are being licensed, the use of waste minimization measures can be required as a condition of the license.

The new waste law will also require recycling of wastes. It establishes recycling as a priority over other types of waste management when technically and economically feasible and if a market is available or can be created for recycled goods. Section 14 of the new law employs two clever tools. It authorizes the federal government to require that certain waste be separated from other waste for the purpose of recycling. Moreover, it specifies that producers or sellers of certain products may be required to accept their return after use. Both these tools can be used to increase the prospects of reuse considerably. This provision also authorizes the ministry to require that certain products must have labeling that specifies the manner in which the product or container should be managed after use and to require that certain packaging be used or that deposits be imposed in order to facilitate recycling of packaging materials.[62] The readiness of the German officials to provide leadership in the creation of waste reuse programs is further described in the chapter on waste oils (see Chapter 6).

One West German state, Hessen, has previously put one of the provisions of the new law to work in the area of waste minimization using a prior law, the Federal Emission Protection Law. Whenever new industrial facilities are proposed, the regional planning authorities and the state pollution control agency require as a condition of approval that a showing be made that all feasible waste reduction and recycling measures will be incorporated into the design. This requirement was successfully applied to a chemical plant on the Main River. Waste hydrochloric acid from a unit producing chlorinated hydrocarbons was discharged into the river until a permit was sought for expansion of the plant. As part of the permit, the recovery of the waste acid was required to be implemented within two years. The firm originally filed a lawsuit challenging the authority of the planning agency to issue such orders, but has since withdrawn its suit and begun implementation of the orders.[63] It is difficult to envision that all state and local planning authorities in the United States would have the technical capability to determine which process modifications would be technically and economically feasible. But many local governments may be able to implement programs to at least place the burden of proof on industrial polluters to demonstrate that they have considered waste reduction and recycling measures.

Finally, the West German UBA is charged with advising the Federal Maritime Agency on the administration of the provisions of the Oslo and London conventions on ocean dumping and ocean incineration of waste. Through the provisions of these treaties, which require signatories to explore the possibilities of land-based utilization and treatment of wastes before ocean dumping and incineration are permitted, the UBA has been requiring the recycling of wastes that were formerly dumped or incinerated at sea.[64] These developments are examined, in full, in Chapter 4.

PUBLIC INITIATIVES: SECURING SAFE SUBSTITUTES THROUGH THE MARKETPLACE

Every consumer of products and services makes countless decisions each week that influence the generation of waste. Many of the products used by consumers are hazardous in and of themselves and become hazardous waste when disposed of. Many more products result in the generation of hazardous waste or toxic pollution in their manufacturing process. In at least one European nation, Austria, there is a small but growing movement of citizens who are exercising their choice as consumers to encourage the use of safer chemical substitutes.

As a deliberate analogy to the "soft energy" movement of the 1970s, the Austrians involved call this movement the "soft chemistry" movement. In response to the consciousness raised by proponents of this concept, a whole market for safe substitutes has arisen, from paints without synthetic organic solvents or heavy metal pigments to alternative household cleaners. Although its impact is certainly small thus far, this movement might yield a fuller consumer movement that could have a major impact on the generation of hazardous waste.[65]

In the United States this type of consumer movement for safer substitutes is growing out of the recognition that hazardous household products are being disposed of by the ton into municipal landfills, where they threaten drinking water supplies. Several of the local government programs to encourage better disposal of household toxics also contain educational programs aimed at encouraging consumers to use less hazardous products. Some of the more successful small generator and household collection programs both in the United States and in Europe are examined in Chapter 10. This greater involvement of citizens in waste issues may begin to help them make choices that significantly reduce waste generation in the future.

LESSONS FROM EUROPE

As noted earlier, several American states have begun waste reduction programs. As they have often done in the 1980s, these state initiatives are serving as laboratories for more widespread innovations in both actual management and proposed policy changes. Following these leads, Congress charged the Environmental Protection Agency with determining what measures should be used to encourage waste reduction on the national level and to issue waste reduction certification requirements under the Resource Conservation and Recovery Act amendments of 1984. With these demands in mind, there are some overall lessons that can be derived from the European experience with pollution prevention.

The first lesson is that pollution prevention needs a larger governmental home than hazardous waste regulatory departments. It must have a high-level, highly visible focus and stable institutional base with clear links to agencies responsible for economic development and technology transfer. By being extended beyond regulatory agencies alone, the task is reconceived as a combination of economic development and environmental protection. The success of institutionalizing waste reduction within European governments, as seen by the examples from France and the Netherlands, depends to a large extent upon the support of the concept by lead agencies dealing with industrial policy and technical innovation. The best institutional home, then, for pollution prevention programs is probably in agencies without direct regulatory authority.

Second, pollution prevention should not be limited only to hazardous waste reduction or to a focus on any one environmental medium. It is inevitably tied to an intergrated management approach and can best be implemented in a cross-media fashion. Nearly all of the European programs reviewed deal with all types of pollution, and some of the programs include energy and water conservation. As noted earlier, European "clean technology" programs are broader still, since they start from the premise that the choice of technology itself should be scrutinized to ensure that effects on environment and public health are minimized. The Norwegian regulatory program appears to be a concrete example of how a focus on pollution prevention in a regulatory program is facilitated by having the program set up across media boundaries, allowing for a focus on overall reduction of pollution from a given process or industry rather than permitting a reduction in one medium that may actually increase pollution in another medium.

Third, pollution prevention should be promoted as an integral part of a nation's overall policy to encourage industrial efficiency, productivity, and innovation. This link between environmental strategies and the revitalization of a nation's industries deserves careful development in the United States. This view of pollution prevention not only helps alleviate the environment versus economy tension, but it also calls for the use of positive measures such as demonstration grants and technology transfer systems. The French and the Dutch have made pollution prevention an important part of their explicit industrial policy and support pollution prevention projects through industrial development and market support pollution prevention projects through industrial development and market research programs. The Dutch also consider clean technologies to be a major component of their technological export program, capitalizing worldwide on their new approaches.

Fourth, widely disseminating information on pollution prevention strategies is a relatively inexpensive but effective way to encourage further reforms by similar industries. The more detail that can be provided with these examples, within the limits of non-proprietary information, the more impact the information has. The French river basin agencies, for instance, are attempting to provide detailed economic information in their pollution prevention database. There are several European strategies for disseminating this information. The most effective ones are those that utilize existing industrial networks such as trade associations and industrial development agencies. Moreover, computer databases and international networks show great promise of making thousands of examples available to industries throughout the world.

Fifth, there is a need for governments to help finance the full-scale demonstration of effective waste reduction strategies. Actual performance statistics are a vital part of the persuasion, and the more direct the government help the more rapid the development. Although many such technologies are in use today, many more have been developed but not demonstrated because of the reluctance of industry to risk capital on new technologies that are not required by regulations and that may be more expensive in the short run than disposal methods that are still available. The coming restrictions on land disposal of hazardous waste will certainly provide a strong incentive for American waste generators to look for technologies that reduce waste generation, and an extensive program of technology demonstration and technology transfer would greatly facilitate that transition. A closer look at European grant selection procedures and subsequent promotion tactics is needed, since significant levels of subsidies for pollution prevention technology demonstration are universally provided in European programs.

Sixth, other financial incentives can be effective, such as low-interest loans, grants for capital expenditures, and tax breaks, but there is concern that such incentives violate the principle that polluting enterprises should pay the full costs of dealing with their pollution in an effort to avoid economic distortions. There are few European programs that actually provide capital assistance in these times of tight budgets. The French river basin agencies provide a good example of how to subsidize capital investments without violating the polluter pays principle, since the funds used to subsidize pollution prevention projects are generated from taxes on pollution. From this perspective, the waste-end tax idea may be even more appealing as a revenue source for incentive programs and technology transfer programs than for dumpsite cleanup and regulatory programs.

Seventh, mandatory measures are effective in encouraging further pollution prevention even when they are not legally enforced. The fact that a government can call upon mandatory authority generally provides a strong incentive for voluntary measures and encourages the use of available programs for information dissemination, technology transfer, and financial assistance. At the heart of mandatory measures adopted in Europe is a simple, but powerful philosophy: there is no inherent right to generate waste, to use processes that result in harmful pollution or to allow the inefficient use of resources harmful to human health or the environment. This explicit denial of the right to pollute adds considerable weight to the waste reduction campaign. As discussed earlier, the state of Hessen in the Federal Republic of Germany has gone the farthest in implementing this philosophy directly.

Eighth, product control programs, such as the Toxic Substances Control Act in the United States, can be used more aggressively to encourage the development and utilization of safer subsititutes that prevent pollution and reduce human exposure to toxic substances. Some European programs have extended restrictions on the production and use of hazardous chemicals to chemicals that have not been controlled in the United States. There should be a stepped up use of the Toxic Substances Control Act in this country.

Finally, informed consumerism can play an increasing role in the prevention of pollution. As more consumers become aware of the toxic hazards of some of the products they use, they might begin to wield their purchasing power to reduce the generation of pollution. Given adequate information and a reasonable choice, many consumers would choose less polluting products, as the "soft chemistry" movement in Austria shows.

E. F. Schumacher, author of *Small Is Beautiful* and developer of the concept of appropriate technology, noted one of the central features of pollution prevention when he wrote: "It is my experience that it is rather more difficult to recapture directness and simplicity than to advance in the direction of ever more sophistication and complexity. Any third-rate engineer or researcher can increase complexity, but it takes a certain flair of real insight to make things simple again." Waste reduction and the more robust quest for clean technologies help simplify the management of risk. Reducing waste avoids transport risks, decreases the need for complex treatment facilities and the delicate political balancing needed to site them and can eliminate the complicated task of defining and regulating the risks of waste disposal. The European example shows us, however, that the establishment of effective waste reduction programs is not a simple task. By incorporating a broad range of policy measures that dissolve barriers born from the lack of sufficient information, capital, and efficient technology, the European example shows clearly that consistent government leadership is needed to make pollution prevention more than just a good idea.

NOTES

1. Low Waste Technology. E.C.E. Report of the Seminar on Low-Waste Technology. Tashkent, U.S.S.R., United Nations Economic Commission for Europe, Geneva, Switzerland, 1984.

    2. United Nations Economic Commission for Europe, "Declaration on Low- and Non-Waste Technology and Reutilization and Recycling of Wastes," Geneva, Switzerland, Feb. 11, 1980.
    3. Frank Van den Akker, Minister of Housing, Physical Planning, and Environment, the Netherlands, "Recycling and Clean Technologies in the Netherlands: An Example for Europe," Leidschendam, Netherlands, 1985.
    4. J. C. D. Fisher, "The Implementation of Cleaner Technologies," paper presented at the Conference on Low- and Non-Waste Technologies: Theory and Practice, Zel Am See, Austria, October 8–9, 1985.
    5. Ibid.
    6. Ibid.
    7. Rudolph Kauders, *Umweltfreundliche Technologien aus Österreich* [Austrian Non-Waste Technologies], Schriftenreihe der Arbeitsgemeinschaft Umweltschutz der Arbeitgeberverbände der Österreichischen Gesellschaft zur Reinhaltung der Luft (Vienna, Austria, 1985).
    8. "Techniques propres dans l'industrie—22 exemples du basin Nord-Artois-Picardie" [Clean Technologies in Industry—22 Examples from the North Artois–Picardy River Basin], Agence de l'Eau Artois-Picardie, 1980.
    9. United Nations Economic Commission for Europe, "Declaration on Low- and Non-Waste Technology and Re-Utilization and Recycling of Wastes."
    10. United Nations Economic Commission for Europe, *Compendium of Low- and Non-Waste Technology,* Geneva, Switzerland, vols. 1–2 (1981), 3–4 (1982), 5 (1983), 6 (1984).
    11. Hungarian National Authority for Environment Protection and Nature Conservation, Compendium of Lectures on Low- and Non-Waste Technology, Budapest, Hungary, United Nations Economic Commission for Europe, Geneva, Switzerland, December 1983.
    12. Anthony Rowland, "Policy for the Development of Clean Technologies in the European Community," Brussels, Belgium, 1985 (unpublished paper).
    13. Van den Akker, "Recycling and Clean Technologies in the Netherlands."
    14. Klaus Muller, Clean Technology Office, Danish National Agency for Environmental Protection, personal communication, August 14, 1985.
    15. Jean-Claude Chazelon, *Les techniques propres dans l'industrie française,* French Ministry of the Environment, Mission for Clean Technologies (Paris, France: La Société Objective, 1982); Michael Overcash, *Techniques for Industrial Pollution Prevention,* Chelsea, Mich.: Lewis Publishers, 1986).
    16. French Ministry of the Environment, Directorate for Prevention of Pollution, Mission for Clean Technologies, "Clean Technologies in France," 1983.
    17. Ibid.
    18. Dutch Ministry of Housing, Physical Planning, and Environment, Department of Wastes and Clean Technology, "Multi-year Program for Chemical Waste (1985–1989)," 1985.
    19. Rowland, "Policy for the Development of Clean Technologies in the European Community."
    20. Reinhard Lagler, "Special Waste in Austria: An Economic, Legal/Institutional Analysis" (Master's thesis, University of Vienna, Vienna, Austria, November 1985).
    21. Kauders, *Umweltfreundliche Technologien aus Österreich.*
    22. Danish Ministry of the Environment, Order no. 532 of October 16, 1984, Promulgating Act on Recycling, Reuse, and Reduction of Waste.

23. Ibid.

24. Klaus Muller, Clean Technology Office, Danish National Agency for Environmental Protection, personal communication, August 14, 1985.

25. Danish National Agency for Environmental Protection, Office of Clean Technologies, "List of Projects Funded as of May 23, 1985," August 1985.

26. Peter Jonsson, Danish National Agency of Environmental Protection, "Danish Legislation and Research and Development Activities in the Field of Cleaner Technology," Copenhagen, Denmark, 1985 (unpublished paper).

27. French Ministry of the Environment, "Clean Technologies in France."

28. Agency for Recycling and Disposal of Waste (ANRED), *Report of Activities,* Angers, France, 1981.

29. Agency for Recycling and Disposal of Waste (ANRED), *Report of Activities,* Angers, France, 1983.

30. French Ministry of the Environment, "Clean Technologies in France."

31. Ibid.

32. Van den Akker, "Recycling and Clean Technologies in the Netherlands."

33. Ibid.

34. Georg Goosman and Hans Sutter, Section for Waste Minimization, West German Federal Environmental Agency (UBA), personal communication, West Berlin, June 26, 1985.

35. Ibid.

36. West German Parliament, "Answer of the Federal Government to the Questions of the Minority Parties Concerning Special Waste Management," October 12, 1983.

37. Swedish National Environmental Protection Board, *Environmental Protection in Sweden* (Solna, Sweden, November 1984) p. 21.

38. University of Lund, TEM Brochure, Sjobo, Sweden.

39. Agency for Recycling and Disposal of Waste (ANRED), *Report of Activities,* Angers, France, 1983.

40. French Ministry of the Environment, "Clean Technologies in France."

41. Georg Goosman, Section for Waste Minimization, West German Federal Environmental Agency, personal communication, West Berlin, June 26, 1985.

42. Jonsson, "Danish Legislation and Research and Development Activities in the Field of Cleaner Technology."

43. Agency for Recycling and Disposal of Waste (ANRED), *Report of Activities,* Angers, France, 1981.

44. French Ministry of the Environment, "Clean Technologies in France."

45. Blair Bower, *Incentives in Water Quality Management: France and the Ruhr Area,* Resources for the Future Research Paper R-24, Washington, D.C., 1981, p. 280.

46. French Ministry of the Environment, "Clean Technologies in France."

47. Ibid.

48. Bjorn Bergmann-Paulsen, Assistant General Director, Norwegian State Pollution Control Authority, personal communication, August 18, 1985.

49. Swedish Institute, "Fact Sheet Environment Protection in Sweden," Stockholm, Sweden, 1978.

50. Ibid.

51. Dutch Chemical Waste Act, February 11, 1976, section 37.

52. Dutch Ministry of Housing, Physical Planning, and Environment, "Multi-year Program for Chemical Waste (1985–1989)."

53. Klaus Muller, Clean Technology Office, Danish National Agency for Environmental Protection, personal communication, August 14, 1985.

54. Danish Ministry of the Environment, Environmental Protection Act, as per January 1, 1983, part 5.

55. Danish National Agency for Environmental Protection, "NAEP—Denmark" (brochure), 1982; Klaus Muller, Clean Technology Office, Danish National Agency for Environmental Protection, personal communication, August 14, 1985.

56. Klaus Muller, Clean Technology Office, Danish National Agency for Environmental Protection, personal communication, August 14, 1985.

57. Michael Despax and William Coulet, *The Law and Practice Relating to Pollution Control in France,* 2d ed. (London: Graham and Trotman, 1982), p. 85; Etienne LeRoy, "Processing Hazardous Wastes in France," United Nations Environment Program Industry and Environment Special Issue, vol. 3, no. 8, (1983), pp. 46–51; M. Grenet, French Ministry of the Environment, Directorate for Pollution Prevention, Division of Industrial Waste, personal communication, June 1985.

58. Dutch Chemical Waste Act (February 11, 1976), sections 33–34.

59. Morten Helle, Norwegian State Pollution Control Authority, "Status of Norwegian Hazardous Waste Management", paper presented at the World Commission on Environment and Development Meeting on Management of Hazardous Waste in Nordic Countries, Oslo, Norway, June 24–25, 1985.

60. Norwegian Ministry of the Environment, Regulations Concerning Delivery, Collection, Treatment, and Disposal of Certain Categories of Hazardous Waste, April 10, 1984.

61. Karin Refsnes, "Control of Toxic Substances in Scandinavia," paper presented at the Conference on Toxic Substances, Human Health, and International Regulations, London, November 13–15, 1984.

62. West German Federal Government, Draft of a 4th Amendment to the Waste Law, Bonn, February 21, 1985; Joanne Linnerooth, International Institute of Applied Systems Analysis, personal communication, May 12, 1987.

63. Carl Otto Zubiller, Director of Waste Management, Hessian Ministry of the Environment, Wiesbaden, Federal Republic of Germany, personal communications, July 3, 1983 and July 11, 1985.

64. Hans Sutter, Section for Waste Minimization, West German Federal Environmental Agency, personal communication, June 26, 1985.

65. Hanswerner Mackwitz and Barbara Koszegi, *Zeitbombe Chemie: Strategien zur Entgiftung unserer Welt* [The Chemical Timebomb: Strategies for Detoxifying Our World] (Vienna, Austria: Verlag ORAC, 1983), pp. 265–70.

3

Shifting the Burden off the Land: The Role of Technical Innovations

GARY DAVIS

INTRODUCTION

Over fifteen years ago several European nations recognized that chemical waste management requires chemical technology. While the household refuse handlers of America were still mainly in charge of hazardous waste management, European facilities more closely resembling chemical manufacturing facilities than dumpsites were employing chemical engineers and technicians to recycle, detoxify, and destroy these wastes. This divergence between European management strategies and those in the United States has attracted the attention of American engineers and policymakers and has provided a major impetus for the movement to restrict land disposal in the United States.

The European lead in hazardous waste management is not, strictly speaking, a lead in technology. Many American firms possess the same technologies, but have had difficulty in putting them into practice because of the continued reliance on land disposal in this country. The European lead is in the implementation of these technologies and in the operating experience that comes from their long use. This lead in implementation was won because of the willingness of European governments to push technology by regulations, financial incentives, and shared ownership of treatment facilities. As American hazardous waste technology is finally being unleashed by changes in the hazardous waste regulatory program, it is important to consider lessons gained from European experience to ensure that we do not move carelessly into this new era of hazardous waste management.

INTEGRATED TREATMENT FACILITIES

European engineers first introduced the concept of the full-service hazardous waste treatment facility. Integrated treatment facilities include all or most of the processes needed to treat the full range of hazardous wastes at the same site. They typically include processes for destroying organic waste, for detoxifying inorganic waste, for treating waste water, and for disposing of the resulting treatment residues.

Including all of these processes at the same facility has obvious technical advantages. In many cases, one waste can be used to treat another, as in the neutralization of acidic waste with caustic waste. Energy recovered from the incineration of organic waste can be used to provide steam and electricity for the operation of other treatment processes. Oily waste separated from oil/water emulsions can be used as fuel to burn other organic waste. The residue from one treatment process may be further treated by another type of treatment process, as in the treatment of incinerator scrubber water by physical/chemical processes used to treat metal-containing waste streams.

Economic advantages are achieved by applying a single administrative and support service (laboratory, recordkeeping, and so on) to as broad a portion of the waste stream as possible. Management advantages are achieved by having the choice of the most appropriate technology for managing a particular waste stream made by the treatment facility operator, who can choose from among the full range of technologies, at the time the waste is received, rather than by the generator or the waste transporter. Finally, integrating all necessary treatment operations at a few large facilities should, in theory, make enforcement and control by regulatory authorities easier, since there would be fewer facilities to police.

There are also disadvantages to integrated treatment facilities that may outweigh their technical, economic, and administrative advantages. The foremost disadvantage is the difficulty in siting large integrated treatment facilities. Since most European integrated treatment facilities were constructed prior to the time when public concern was focused on the hazardous waste issue, public opposition was not significant. But recently constructed facilities in Europe have encountered stiff opposition, and American waste management companies have had little success in siting large integrated facilities. The principle issue, as discussed in more detail in Chapter 8, is the perceived inequity of importing large quantities of hazardous waste from outside the facility area. A further disadvantage of centralizing waste management is that longer transport distances would typically be required for the waste with attendant increases in the risks of hazardous waste transportation.

Despite these disadvantages, American firms, believing the economic and administrative advantages to be significant, continue to attempt the siting of integrated treatment facilities. State governments, such as those of New Jersey, New York, and Arizona, have generally attempted to facilitate their siting.

Table 3.1
Amounts of Hazardous Waste Treated at Kommunekemi in 1983

Oily Waste	21,000 tons
Halogenated Organics	1,700 tons
Nonhalogenated Organics	28,500 tons
Pesticide Wastes	100 tons
Inorganic Wastes and Solutions	9,000 tons
Miscellaneous (includes pharmaceuticals, empty containers)	5,000 tons
Total	65,300 tons

Kommunekemi: A Model Integrated Treatment Facility

The Danish Kommunekemi facility near the town of Nyborg represents a model integrated treatment facility that has been operating since 1975. The Danish hazardous waste collection system, which is described in more detail in chapter 10, funnels nearly all of the nation's hazardous waste to this integrated treatment facility, located as close as possible to the geographical center of this Maryland-sized nation. The 21 transfer stations in this collection system are situated throughout the country so that no waste generator has more than a 30–mile transportation distance to the nearest one.[1] This well-designed collection system helps mitigate the increased transportation risks caused by centralization.

The Kommunekemi facility, which treats approximately 65,000 tons per year of hazardous waste and waste oil, consists of an incineration plant, a waste oil recovery plant, and a physical/chemical treatment unit. A landfill for treatment residues is located about twelve miles away from the facility.[2] Table 3.1 is a breakdown of the types and quantities of waste that were treated at the facility in 1983.[3]

The incineration plant at Kommunekemi consists of two large rotary kiln incineration systems (rotary kiln incineration is described in more detail later), one which has been operating since 1975 and the other since 1982. Their combined capacity is approximately 90,000 tons of waste per year, which can be burned in solid, liquid, or sludgy form (including waste in drums).[4] Heat from incineration is recovered in a waste heat boiler, piping approximately 35 tons of steam per hour into the local district heating system. This provides about 35 percent of the total heat demand of the town of Nyborg.[5]

The first incinerator uses a dry scrubber and an electrostatic precipitator for air pollution control, resulting in a dry residue that is sent to the landfill together with the ash from the incinerator. The second incinerator uses a wet/dry scrubber (described in more detail below) and an electrostatic precipitator for air pollution control, and the residues are also sent to the landfill for disposal. The volume of dust and ash from incineration for disposal is approximately 20 percent of the volume of the waste before it was burned in the incinerator.[6]

The waste oil recovery plant at Kommunekemi has a capacity of about 18,000 tons per year of all types of oily waste including sludges and oil/water mixtures. Heating the waste oil mixture to approximately 90° C causes the volatile substances (such as gasoline) to evaporate and the rest of the waste oil to separate into a sludge layer and an oil/water mixture. The volatile substances are condensed and incinerated, and the oily sludge is collected and incinerated with other sludge waste. The oil/water mixture is further processed to separate the two phases. The oil is used for fuel for the incinerators and the water is injected into the afterburners of the incinerators to cool the hot gases before they enter the boiler and air pollution control system.[7]

The physical/chemical treatment plant treats inorganic waste and solutions, which generally fall into four categories:

1. Alkaline solutions containing cyanides
2. Acid solutions containing chromium
3. Acid solutions containing iron (pickling liquors)
4. Other solutions, including those containing mercury and fluorides, that cannot be treated with the first three categories

The plant employs a stepwise process for treating the first three categories of waste in the same reactor vessel using one type of waste to treat others. The cyanides are destroyed by treatment at high pH with sodium hypochlorite; hexavalent chromium is reduced to less toxic trivalent chromium by mixing iron-containing acidic waste into the reactor; and the toxic metals are precipitated from solution by addition of lime. The resulting sludge is dewatered in a filter press and disposed of in the landfill. The aqueous portion is analyzed for contaminants and treated further, if necessary, before discharge into the sewage treatment system for the town of Nyborg.[8]

Aqueous waste containing mercury, such as waste from analytical laboratories, cannot be adequately treated using hydroxide precipitation. As a result, the Kommunekemi facility employs sulfide precipitation for these solutions, and the filtered sludges are placed in a special area of the landfill. Finally, to prevent generation of hydrogen fluoride, fluoride solutions are treated separately by lime precipitation.[9]

The residues from the incineration and treatment processes, which total about 18,000 tons per year, are sent to the landfill, which also receives a very small amount (less than 1,000 tons per year) of solid chemical waste that is suitable for land disposal without treatment. This includes solid materials with low organic chemical content and with low concentrations of soluble inorganic chemicals.[10]

The landfill is located near the sea, and the direction of groundwater flow in the area is toward the sea, lessening the chances of leachate from the landfill contaminating drinking water supplies. The landfill cells, which are separate for different types of waste (metal sludge filter cakes, incinerator ash, other solid waste), are lined with plastic membranes and have an underdrain system and monitoring wells. The cells for disposal of metal sludge filter cakes are also filled with lime to maintain a high pH to prevent metals from leaching and are covered with a plastic membrane each day. The final cover for the cells is around 1.5 feet of clay and 1 foot of topsoil.[11]

According to the Danish National Environmental Protection Agency, which monitors the Kommunekemi facility through a resident inspector stationed at the site, the environmental track record of the facility is good. There were odor problems in the early years of operation, which were remedied by shutting down an older liquid incinerator and replacing it with a new rotary kiln and by putting air pollution controls on organic liquid storage areas.[12] There have been only a few relatively minor releases of hazardous or noxious substances at the site in its ten years of operation. These included an incident of puffing of flue gas from a rotary kiln afterburner, accidental release of part of the contents of a storage tank of solvent waste into the sea, and the spill of an odorous waste in the receiving area that spread fumes into the community.[13]

Other Integrated Treatment Facilities

There are several other integrated treatment facilities operating in Europe with varying combinations of treatment processes. Table 3.2 is a listing of these facilities. For a more detailed description of these facilities, see Appendix 1.

INCINERATION IN EUROPE

While the concern about incineration is growing in Europe, European nations rely to a much larger extent on incineration of both hazardous and municipal waste than does the United States. As the United States begins to implement land disposal restrictions for many types of organic hazardous waste and as our municipal landfills begin to fill up, policymakers need to take a long look at the European experience with incineration to determine the role of incineration in this nation's waste management future.

There are several hazardous waste incinerators operating in the United States today, and several more are proposed. For municipal solid waste there is a boom in the construction of mass-burn plants to replace local landfills. As the United States turns to incineration for waste disposal, there is a growing concern about the air emissions from these incinerators and whether we are trading one type of hazard (groundwater pollution) for another of equal or greater magnitude (toxic air emissions).[14] This concern is heightened by evidence that the operating

Table 3.2
Integrated Hazardous Waste Treatment Facilities in Europe

Country	Location	Owner	Capacity	Technologies	Reference
Austria	Vienna	EBS	100,000	Incin., phys./chem.	11
Finland	Riihimaki	Oy Suomen Ongelmajate	71,000	Incin., phys./chem., landfill	12
France	Mitry-Compans	GEREP	45,000	Incin., phys./chem.	13
France	Limay	SARP	98,000	Incin., phys./chem., solidification	13
Sweden	Norrtorp	SAKAB	60,000	Incin.,phys./chem., landfill	14
West Germany	Schwabach	ZVSMM	120,000	Incin., phys./chem., landfill	15
West Germany	Ebenhausen	GSB	150,000	Incin., phys./chem., landfill	16

Sources:

1. Herbert Hofstetter, Plant Mananger, EBS, Personal communication, September 1982.
2. F. Aarnio, "The Finnish System," paper presented at the First International Symposium on Operating European Centralized Hazardous (Chemical) Waste Management Facilities, Odense, Denmark, September 1982.
3. *Guide pour l'élimination et la Valorisation des déchets industriels* [Guide for Disposal and Recycling of Industrial Waste], French Ministry of the Environment, 1984.
4. "This is SAKAB," SAKAB, Norrtorp, Sweden, 1984.
5. "Zweckerband Sondermüllplatze Mittelfranken: German Experiences with the Disposal of Special Wastes," ZVSMM, Schwabach, 1982.
6. "Disposal of Special Waste in Bavaria," Gesellschaft zur Beseitigung von Sondermüll in Bayern GmbH, Munich, FRG, 1983.

track record of hazardous waste incineration facilities in this country has been poor.[15]

The European nations that have long experience with high-temperature hazardous waste incineration consider risks to the public to be low, but express concerns over the emissions of products of incomplete combustion of chlorinated organics (principally tetrachlorodibenzo-p-dioxin and dibenzofuran compounds) and emissions of heavy metals, such as lead, mercury, and cadmium. The European response to these concerns has been to pay careful attention to the types of waste that are fed into the incinerators and to the maintenance of optimum operating conditions that have been established from years of experience. Stringent air emissions limits, including limits on heavy metal emissions, have necessitated the use of innovative and highly efficient air pollution control technologies.

Nonetheless, monitoring at some European hazardous waste incinerators has shown the presence of low levels of dioxins, and few incinerators are permitted to burn highly chlorinated organic wastes, such as polychlorinated biphenyls (PCBs). As for heavy metal emissions, European incinerators generally perform better than their American counterparts, which do not usually have limitations in their permits on heavy metal emissions.[16] At one incineration facility in West Germany, however, monitoring of nearby fields for heavy metals showed elevated levels of lead, mercury, and cadmium, necessitating plans to install improved pollution control equipment.[17]

Although municipal waste incinerators are much more widely used throughout Western Europe than in the United States, it is important to note that the growing concern over dioxin emissions from mass-burn plants has caused at least two governments, those of Sweden and of the state of Hessen, West Germany, to postpone consideration of any new facilities.[18] Furthermore, concerns over heavy metal emissions from these facilities have led to programs in several countries to collect and recycle small quantities of household and commercial waste containing heavy metals, such as mercury-cadmium batteries, in order to keep them out of the burn plants (see Chapter 10).

The Advent of the Rotary Kiln Incinerator

Rotary kiln incineration for hazardous waste destruction was first developed in the West German chemical industry in the 1960s by companies such as BASF and Bayer.[19] Rotary kiln–type burners had previously been used in combusting municipal garbage, in production of lime, and in cement production, but had not been adapted for chemical waste destruction. The most significant changes were the addition of a secondary combustion chamber and the development of high-temperature fire brick enabling the maintenance of the over 1,200° C temperatures required for destruction of toxic organic compounds.

The first rotary kiln incinerators designed specifically for hazardous waste destruction were constructed by BASF at its Ludwigshafen, West Germany, chemical plant in 1964. Two kilns with secondary combustion chambers and waste heat boilers were constructed at the plant to burn approximately 25,000 tons per year of solid, liquid, and sludgy waste. Two more rotary kiln incinerators were built at the BASF plant in 1969 with improvements in design and operating conditions learned from the first two kilns. These kilns were also constructed to have nearly twice the capacity of the first two. By 1970 over 80,000 tons per year of hazardous wastes were being incinerated at the BASF plant.[20]

Today the rotary kiln incinerator is the workhorse of most European hazardous waste treatment plants. This type of incineration system can destroy organic waste at temperatures over 1200° C with greater than 99.99 percent efficiency. These incinerators have four elements as shown in Figure 3.1. The kiln, a slowly rotating metal cylinder lined with fire brick, burns solids, sludges, and liquid waste directly or in containers. The kiln is often as much as 10 feet in diameter and more than 40 feet long. An afterburner is designed to complete combustion of gases from the kiln and can be used to burn liquids directly. Most European incinerators next incorporate a heat recovery system to recover energy from the hot exhaust gases to generate steam and electricity. Following the heat recovery system, an air pollution control system removes particulates and acid gases from the exhaust gases.

In order to meet air pollution standards, European incinerators have been designed with air pollution control systems that have become increasingly sophisticated. The principle conventional pollutants emitted from hazardous waste incinerators are particulates and hydrochloric acid gases. The particulates often contain heavy metals from sludges and other inorganic wastes that are burned. The hydrochloric acid gases are generated from the combustion of chlorinated organic compounds.

Although rotary kilns are capable of burning waste in practically any form, European incinerator operators are generally very careful about the types and amounts of wastes that are fed into the kilns. Because of the concern for emissions of heavy metals, the Danish Kommunekemi facility, for instance, attempts to restrict the introduction of relatively volatile metals, such as mercury, cadmium, and lead, into the kilns.[21] Air pollution permits for hazardous waste incinerators in Europe are generally written with limits for heavy metal emissions as well as particulates and HCl. Table 3.3 is a compilation of permit levels for European incinerators with American requirements presented for comparison.

In order to keep levels of hydrochloric acid in the combustion gases to the required level, incinerator operators first control the concentration of chlorine in the feed. For incinerators with the most effective scrubbers, such as the HIM incinerator near Biebesheim, West Germany, chlorine contents up to 10 percent are generally acceptable in the feed to the kiln.[22] Thus, only those incinerators with the most effective scrubber systems are permitted to burn highly chlorinated compounds such as PCBs.

The most effective scrubbing systems used in European incinerators are those that use a liquid to scrub the acid gases from the flue gas. Facilities with dry scrubbers, such as the EBS facility in Vienna, are not able to burn highly

Figure 3.1
Rotary Kiln Incineration System

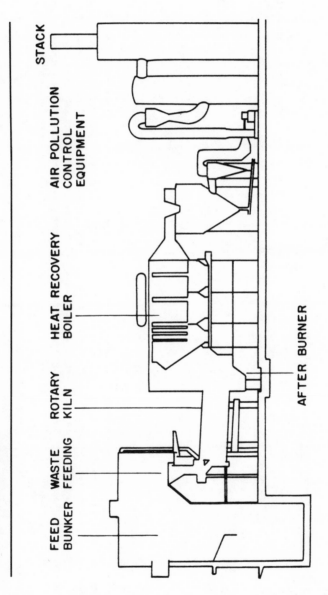

FEED BUNKER   WASTE FEEDING   ROTARY KILN   HEAT RECOVERY BOILER   AIR POLLUTION CONTROL EQUIPMENT   STACK

AFTER BURNER

Table 3.3
Permit Limits for Hazardous Waste Incinerators in Europe

Component	Komunekemi[a]	ZVSMM[b]	HIM[c]	GSB[d]	SAKAB[e]	EBS[f]	RCRA[g]
			Concentration in Stack Gas (mg/Nm^3)				
HCL	300	50	100	100	35	100	41b/hr. or 99%
HF	5	2	5	5	5	5	
SO_2	750	200	200	400		300	
Cl_2	25						
NO_x	300						
Dust	100	500	75	50	35		180
Organics		50 / 20	50				99.99% DRE
As	0.1						
Hg						1	
Cd			0.16			0.1	
Cr	5					0.1	
Pb						1	
Zn						5	
Be			0.1			5	
(As,Hg,Cd)	0.02 in sum						
(Cr,Ni,Th,Vd,Pd)	2 in sum						
(As,Pb,Cr,Co,Ni)	1 in sum						

a. Kommunekemi, Nyborg, Denmark, Incinerator III; Arne Kristensen, "Operating the Rotary Kiln Incinerators at Kommunekemi," paper presented at the 2nd International Symposium on Operating European Centralized Hazardous (Chemical) Waste Management Facilities, Odense, Denmark, September 1984.

b. ZVSMM, Schwabach, FRG; personal communication with Dr. Norbert Amsoneit, Chief Chemist, ZVSMM, Schwabach, FRG, July 8, 1985.

c. HIM, Biebesheim, FRG; Gunter Erbach, "Experiences with Special Waste Reception, Intermediate Storage, and Incineration at the Hazardous Waste Incineration Plant at Biebesheim," in U.S. EPA, Hazardous Waste Engineering Research Laboratory, *Proceedings: International Conference on New Frontiers for Hazardous Waste Management*, EPA/600/9-85/025 (Cincinnati, Ohio, September 1985), p. 236.

d. GSB, Ebenhausen, FRG; Franz Defregger, "The Bavarian System for Special Waste Management—15 Years Experience in Collection, Treatment, Disposal, and Control," in U.S. EPA, Hazardous Waste Research Laboratories, *Proceedings: International Conference on New Frontiers for Hazardous Waste Management*, EPA/600/9-85/025 (Cincinnati, Ohio, September 1985), p. 86.

e. SAKAB, Norrtorp, Sweden; "This is Sakab," SAKAB, Norrtorp, Sweden (1984).

f. EBS, Vienna, Austria; Willibald Lutz and Friedrich Habl, "Hazardous Waste Collecting and Treatment in Austria," in U.S. EPA, Hazardous Waste Engineering Research Laboratory, *Proceedings: International Conference on New Frontiers for Hazardous Waste Management*, EPA/600/9-85/025 (Cincinnati, Ohio, September 1985), p. 93.

g. Resource Conservation and Recovery Act regulations, Title 40, Code of Federal Regulations, Section 264.343.

chlorinated organics.[23] Wet scrubbers generally use large quantities of an alkaline solution to neutralize the acid gases and to remove some of the particulate emissions. The solution is usually recycled through the scrubber with the addition of fresh chemicals as needed. Some of the scrubber liquid must be drawn off and discharged to prevent the buildup of salts and solids in the system. This scrubber liquid contains high levels of heavy metals and must be treated before discharge, resulting in a large quantity of sludge for disposal.

An innovative gas cleaning approach has been implemented on some of the newer European incinerators. The actual scrubber is a spray dryer that removes the pollutants and produces a dry residue rather than a scrubber liquid for disposal. The dry residue can be disposed of directly in a hazardous waste landfill or, as with the HIM incinerator in Hessen, West Germany, taken to an underground storage facility in the salt mines of Herfa-Neurode.[24] There are at least four commercial incineration facilities that are equipped with this type of scrubber system: the Kommunekemi facility in Denmark, the SAKAB facility in Sweden, the Ongelmajate facility in Finland, and the HIM facility in Hessen, West Germany.

In this type of system, the flue gases from the incinerator are passed into a reactor where small droplets of the scrubbing liquid are sprayed. The scrubbing liquid neutralizes the acid gases, condenses heavy metals, and collects particulates as the water evaporates away. The dry powder that remains is removed from the bottom of the reactor and from a subsequent dust collector and some of it is recycled to the liquid phase since it still contains some reagent for neutralizing the acid gases. The flue gases then pass on to particulate removal equipment, such as electrostatic precipitators or baghouses, which are able to work more efficiently because of the lower temperature of the gases.[25] The HIM incinerator uses a wet scrubber for particulate removal as well, which is tied in with the spray dryer, so that the only effluent is the dry powder.[26] Figure 3.2 is a diagram of the Niro Atomizer system that is used by the Danish, the Swedish, and the Finnish hazardous waste incinerators.

Cement Kilns for Hazardous Waste Incineration

The Norwegians have been in the forefront of the development of cement production kilns as highly efficient incinerators for organic hazardous waste. The typical cement kiln is a very large rotary kiln operating at temperatures greater than 1400° C in order to produce cement clinker from the combination of limestone, sand, and clay that is fed into the kiln. Rather than financing and constructing a new centralized incineration facility at a price in excess of $30 million, the government of Norway turned to an existing cement plant for the incineration of the nation's organic hazardous waste.

In 1980 the government and Norcem, a commercial cement producer, began a joint program to test the use of a cement kiln for the destruction of hazardous waste. The plant at which the process was tested is located in Slemmestad,

Figure 3.2
The Niro Atomizer Wet/Dry Scrubber

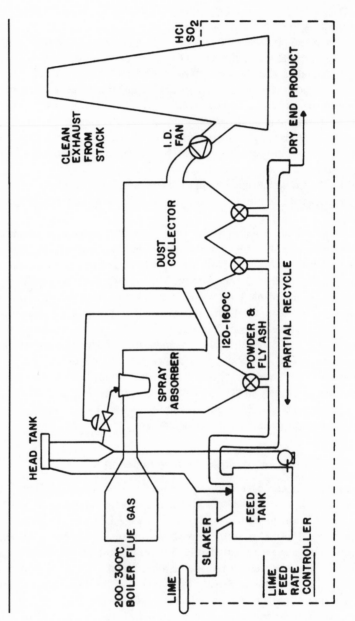

Norway, about twenty miles from Oslo. The kiln that was used for destroying hazardous wastes is 170 meters long (about 200 yards or two football fields) and has a diameter of 5 meters (about 15 feet). It is a wet kiln, which heats the raw materials for cement in a water slurry to produce cement clinker. In order to produce cement, these kilns use tremendous amounts of energy. The Slemmestad plant used 92 tons of fuel oil for the production of 850 tons of clinker per day. Dry kilns, which feed the raw materials in dry form, require less energy for the production of cement clinker and, as a result, are replacing wet kilns.[27]

A hazardous waste feed and blending facility was constructed for the test, and a special nozzle was developed for injecting the waste into the kiln. The nozzle is capable of injecting waste that contains small amounts of solids. During the testing period, hard-to-burn waste such as PCBs and polynuclear aromatic hydrocarbon (PAH) tars from natural gas cracking were burned, and extensive monitoring was performed. The results showed that high levels of destruction were obtained in the kiln (99.99999 percent for PAH tars and 99.99997 percent for PCBs). Low levels of combustion products were found in the stack that represented an increase over reference conditions (fuel oil alone); but significantly, when PCBs were burned, no dioxins were detected at a minimum detection level of 0.5 nanograms per cubic meter. Moreover, no organics were found in the clinker or in the dust from the air pollution control equipment. Tables 3.4 and 3.5 are a summary of the results of the trial burns.[28]

Norcem and the government of Norway are now constructing a full-scale hazardous waste storage and feed system for a dry cement kiln owned by Norcem in southern Norway.[29] They have also jointly performed a trial burn with a dry kiln burning both PAH tars and PCBs, and the results were similar. Destruction efficiencies were 99.9999 percent for PCBs, but 2,3,7,8–TCDD (tetrachlorodibenzodioxin) was found in stack gas at a concentration of 0.1 nanogram per cubic meter.[30]

The combustion of hazardous wastes in a cement kiln has several advantages. A cement kiln is an excellent incinerator because of the high temperatures and long residence times (measured in minutes, not seconds) that must be maintained in order to produce cement clinker. Furthermore, expensive equipment to remove acid gases is not required when burning halogenated organics because the huge mass of alkaline cement clinker absorbs and neutralizes the gases. Because of the high energy requirements of cement production, the combustion of hazardous wastes with a high calorific value, such as waste oil or solvents, results in significant fuel savings. Norcem has estimated that annual fuel savings for its test kiln would be over 10,000 tons of oil. Finally, the kiln operator can charge a fee for the destruction of hazardous wastes, which will add to the economics of cement production.[31]

Because of these advantages the Norwegian government is planning to use cement kiln incineration for the major portion of its organic hazardous waste. Another European nation, France, is making use of cement kilns for hazardous waste incineration. The French reportedly have three cement kilns in operation

Table 3.4
Results from Test Burns at Norcem Wet Process Cement Kiln (PAH Tars)

Burning PAH Tar	Fuel oil & tar		Fuel oil & tar	
	ug/Nm³	g/h	ug/Nm³	g/h
Total organic carbon	40500	3900	8700-9300	840-850
C²-Alkylbenzenes	105	10.2	--	--
C³-Alkylbenzenes	80	7.76	--	--
C⁴-Alkylbenzenes	34	3.3	1	0.99
Alkanes	150	15.3	14	1.27
Benzene	200	20	450	40
PAH	14	1.36	14	1.27
Thiopenes	8	0.775	7	0.635
Keytones	23	2.23	6	0.545
Aldehydes	96	9.31	4	0.365
Phenol	20	1.94	--	--
Organic Acids	30	2.91	--	--
Alcohols	15	1.46	--	--

(--: not measured)

Table 3.5
Results from Test Burns at Norcem Wet Process Cement Kiln (PCBs)

Burning PCB's (48% Cl)	TEST I		TEST II		TEST III	
	ug/Nm3	g/h	ug/Nm3	g/h	ug/Nm3	g/h
C^2-Alkylbenzenes	23	2.3	23	2.3	--	--
C^3-Alkylbenzenes	7	0.7	5	0.5	--	--
C^4-Alkylbenzenes	8	0.8	8	0.8	1	0.1
Alkanes	48	4.8	54	5.4	14	1.3
Benzene	100	10	100	10	450	40
PAH	2	0.2	0.2	0.2	14	1.3
Thiopenes	1	0.1	1	0.1	7	0.64
Ketones	4	0.4	7	0.7	6	0.55
Aldehydes	6	0.6	22	2.2	4	0.37
Chloroform	0.87	87	1.18	118	0.00001	--
Carbon tetrachloride	0.40	40	1.22	122	0.00001	--
Trichlorethylene	0.30	30	0.91	91	0.00001	--
Tetrachlorethylene	0.40	40	1.21	121	0.00003	--
Chlorbenzenes	2.90	290	9.30	930	0.890	89
PCB	0.08	8	0.17	17	0.040	4
Total organic chlorine	2.90	290	--	--	1.6	145

burning about 20,000 tons of hazardous waste per year.[32] In the United States there are as many as six cement kilns burning solvent waste as fuel.[33]

RECYCLING AND PHYSICAL/CHEMICAL TREATMENT TECHNOLOGIES

Hazardous waste recycling and treatment processes used in European hazardous waste facilities have been adapted from those of the chemical and petroleum industries. For instance, as in the United States, solvent recycling by distillation is widely practiced. Standard physical/chemical treatment processes are widely used for the treatment of inorganic hazardous wastes containing cyanides and heavy metals, and waste oils are recovered using a variety of well-proven physical and chemical emulsion-breaking techniques.[34]

Many of the integrated hazardous waste treatment facilities surveyed do not incorporate solvent recycling, since it is a profitable venture practiced by private firms without the need for government subsidization or involvement. Most do accept waste solvents that cannot be readily recovered for reuse as a source of fuel for rotary kiln incinerators. The same situation exists for waste oils. In general, waste motor oils are dealt with separately from hazardous wastes, and are recycled by commercial firms (see Chapter 6).

Oily wastes from refineries and industrial oils and oil/water emulsions are usually managed at hazardous waste facilities and are an important source of fuel for rotary kiln incinerators. Emulsion-breaking processes vary from the thermal process used by the Danish Kommunekemi facility and the Swedish SAKAB facility to the chemical processes used by the GSB and ZVSMM facilities in Bavaria, West Germany. The thermal process used by the Kommunekemi facility was described earlier. In contrast, the chemical process used by the ZVSMM facility in Schwabach, West Germany, uses an iron chloride solution to break the emulsion, followed by filtration of the oily sludge. Because of the high chemical requirements and the necessity to dispose of large quantities of sludge, ZVSMM is now planning an ultrafiltration unit that makes use of membrane separation for the oil/water emulsions.[35]

Treatment processes for heavy metal removal and cyanide destruction are fairly consistent from facility to facility. Cyanide destruction usually involves chemical oxidation of the cyanide ion by a sodium hypochlorite solution to the cyanate or all the way to carbon dioxide and nitrogen. Most facilities also reduce hexavalent chromium to the less toxic trivalent state, using either iron sulfate wastes (Kommunekemi) or sodium bisulfite (ZVSMM). Heavy metals are then precipitated by addition of lime or sodium hydroxide. The Danish Kommunekemi facility is able to save on the purchase of chemicals for treatment by using one waste stream to treat another.[36]

The major disadvantage of these conventional physical/chemical treatment processes is that they produce large volumes of sludges for disposal. Since these are mixed metal hydroxide sludges, they are also presently uneconomical to recover. Because of the difficulties in recovering metals from waste stream mixtures at integrated treatment facilities, the focus of metal recovery technologies is at the source of generation. There are several technologies that show promise, including ion exchange and electrolytic methods. A French company, Dornier, for instance, has established a successful business of installing ion-exchange units in plants that generate metal-containing waste streams and providing central regeneration facilities for regenerating the ion-exchange resins.[37]

IMPROVEMENTS IN DISPOSAL OF TREATMENT RESIDUES

The Germans, known for their reliance on alternatives to land disposal, actually originated the "engineered" or "secure landfill" almost twenty years ago. The original design has only recently become standard practice in the United States with the 1984 RCRA amendments. The Germans also recognized early the limitations of land disposal and have enacted severe restrictions on the types of wastes that may be disposed of in the land. In a continual search for more secure residual disposal, several European nations have also pioneered the use of improved disposal techniques for treatment residues.

The Development of the "Secure Landfill"

Long before other governments in the world made the connection between uncontrolled hazardous waste disposal and groundwater contamination, the government of the district of Middle Franconia in the West German state of Bavaria designed a disposal facility for hazardous waste with special features to prevent groundwater contamination. In 1966, when the Zweckverband Sondermüllplatze Mittelfranken (ZVSMM) was created by the district government, most hazardous waste was either dumped into rivers and streams or disposed of with municipal garbage in what were basically unlined, open dumps.

The future-thinking mayor of the city of Furth, the chairman of an economic development group of local politicians in the district, used political pressure and the fear of drinking water contamination to convince his fellow politicians to create ZVSMM and to require all industrial waste generators within the region to deliver their waste to the facility (see Chapter 5 for more details of the system). He hired a local civil engineer with no prior experience in hazardous waste landfill design to design what was probably the first "secure landfill."[38]

The site was chosen for its natural clay deposits, which were enhanced with an additional clay liner. An underdrain system was constructed to capture any liquid leaching through the landfill, which was then pumped to the surface and treated prior to release into the local sewage treatment system. Hazardous waste was buried in cells, which were covered over with clay and planted over with vegetation to minimize the intrusion of precipitation. Monitoring wells were installed to determine if hazardous components were migrating from the disposal cells.

At first, any and all types of industrial waste were disposed of in this landfill. But after one year's experience, the operators of the Schwabach facility concluded that liquids should not be disposed of in the landfill and that treatment processes were needed to detoxify and render waste less likely to migrate into groundwater. This was not prompted by leaks from the landfill, but by an overall concern about leachate migration. Treatment processes were constructed between 1968 and 1974, and today few untreated hazardous wastes are placed in the ZVSMM landfill. Officials at the facility state that no leachate migration has been detected in monitoring wells around the site, although some suspicion has been created by recent small increases in salts in one of the downgradient wells.[39] Several other European governments have followed the ZVSMM example and have used secure landfills only for treatment residues and other relatively non-toxic, immobile hazardous waste.

Today in the United States, lined landfills with leachate collection systems are being required for the first time as the minimum technology for disposal of hazardous waste. Liquids have only recently been restricted from land disposal, and the Environmental Protection Agency is only now embarking upon a program to restrict the land disposal of certain waste streams that should not be disposed of without pretreatment.[40]

Making Land Disposal More Secure

Above-Ground Disposal

In order to provide added protection to groundwater resources, some European hazardous waste management facilities have been building residual repositories above ground. Probably the first such facility was the Gallenbach landfill constructed by the GSB starting in 1970. GSB, the Gesellschaft zur Beseitigung von Sondermüll in Bayern, is the quasi–publicly owned Bavarian hazardous waste management company that operates the Ebenhausen integrated treatment facility. The landfill is built in a cut into a hillside on top of a clay base. Wastes containing soluble heavy metals are disposed of in drums and covered with concrete, and a plastic cover is put on top of successive layers to prevent infiltration. Leachate is collected at the clay base and flows by gravity into a basin at the foot of the landfill, from where it is pumped into vacuum trucks and transported to the Ebenhausen facility for treatment.[41] Building the landfill up from the surface instead of digging it down into the ground maintains a greater distance between the hazardous residues and the groundwater table and allows the leachate collection system to work by gravity so that it will never suffer from mechanical failure.

The Swedes have also begun using above-ground landfill cells for the disposal of treatment residues at their SAKAB integrated treatment facility near Norrtorp in southern Sweden. Treatment residues are placed on top of layers of clay underlain with a leachate collection system. Great care is taken during the construction of each cell to prevent the intrusion of precipitation by covering the cell with a corrugated metal roof. Each cell is fairly small, also providing less chance for intrusion of water. When the cell is full, a clay cover is placed over the waste and the corrugated roof is moved to the next cell.[42]

Another place where the above-ground landfill design has recently gained favor is in the state of Hessen, West Germany. At the insistence of the Green Party, the political arm of the West German environmental movement, which has recently become part of the coalition state government, the publicly owned state hazardous waste management firm, Hessische Industriemüll (HIM), has been investigating above-ground disposal as an alternative to a proposed in-ground landfill that generated tremendous public resistance. Ironically, the Hessians have turned to an American firm, Rollins Environmental Services, for the technology.[43] Rollins developed the above-ground landfill concept in the United States, but has been unable to put it into wide practice because U.S. regulations still favor more traditional landfilling practices.[44]

Concrete Bunkers for Long-Term Storage

With a high water table and practically all of the surface of the country below sea level, the Dutch have abandoned the traditional concept of land disposal in order to protect their groundwater resources. Although land disposal of hazardous waste took place in the past, the Dutch hazardous waste law now prohibits the practice with very few exceptions. Since the Dutch lack treatment capacity, this has meant that much of the hazardous waste that has been generated in Holland has been exported to other countries.[45] Recently, however, the Dutch have begun developing additional treatment capacity and a new form of long-term storage for treatment residues. The Dutch firm AVR-Chemie, located in Rotterdam, now has in process the construction of concrete bunkers for the long-term storage of hazardous waste treatment residues.[46] (Please see appendix 1 for contacts and references on above-ground units.)

The Salt Mines of Herfa-Neurode

The Europeans have also decided that certain types of hazardous waste are too difficult or dangerous to treat or dispose of where the possibility of environmental contamination exists. To manage these types of waste, the Kali and Salz Company has operated an underground long-term storage facility in the mined-out portion of a salt deposit in the West German state of Hessen since 1972. The stated purpose of the facility is to remove some of the most toxic industrial waste from the biosphere for geological time. Containerized waste is placed in mined rooms in the 300-meter-thick salt formation at a depth of about 700 meters. The deposit is covered with clay and shale layers. The rooms are sealed with brick, but the wastes can still be retrieved for recovery or treatment in the future.

Approximately 35,000 tons per year of waste, such as cyanide salts, mercury sludges, chlorinated pesticides, and concentrated PCBs, are placed into the formation. No liquids or ignitable, explosive, or radioactive wastes are permitted. The greatest single group of wastes that have been disposed of during the operation of the disposal site is hardening salts from the metal-finishing industry. These contain high concentrations of cyanide, making them difficult to treat, and are very water soluble, making them inappropriate for landfilling. Most of the waste comes from West Germany, but a small portion has been accepted from other countries, including the United States. Over 1,000 tons of hazardous waste have been retrieved from the salt mine disposal site for detoxification or recycling.[47]

CONCLUSIONS

Technological innovation does not happen in a vacuum. It is usually spurred by market forces, although it may be spurred by public need translated into government initiative. By deciding early that land disposal was not the optimum hazardous waste solution, European governments spurred the development and implementation of a generation of recycling and treatment technologies that are now being pursued as the means to move beyond dumping in the United States.

The mere fact that these technologies have been used in Europe for nearly twenty years provided compelling evidence for American policymakers that alternatives to land disposal were feasible. This fact significantly aided the shift away from land disposal in the United States, beginning with the California land disposal restrictions.

The most recent innovations in European technology are in the field of waste reduction and clean technologies. As described in Chapter 2, European nations have gone to great lengths to encourage the development of new technologies for waste reduction. This chapter has focused chiefly on treatment technologies, since the United States is still in the early stages of building a treatment infrastructure for hazardous waste that will be restricted from land disposal in the next five years.

The recycling and treatment technologies that are the most widely used in Europe are relatively simple adaptations of standard manufacturing processes. The engineering innovations came in building flexibility into these processes so that a widely variable waste stream could be treated. The rotary kiln incineration system is a good example of that flexibility, since organic waste in any physical form can be destroyed, including waste in containers. The need for this flexibility was heightened by the centralization of treatment facilities in Europe and the fact that the waste stream has gotten "dirtier" as generators have reclaimed more and more of the reusable components.

Enhancing the flexibility built into the technologies, the integration of several treatment processes in one facility provides definite technical and economic advantages. The question of centralization and appropriate scale, however, needs further attention, since large, centralized facilities are difficult to site and increase the overall risks in the transportation of hazardous waste. Furthermore, the presence of centralized facilities that mix waste together from a variety of sources may hinder the preferred alternatives of waste reduction and recycling.

Although several European nations own or heavily subsidize centralized treatment facilities, their governments are not afraid to question and continually upgrade the technologies used. This can be seen in the current debate about incineration in West Germany and Sweden, for instance, and in the care with which incinerators are operated. The limitation on materials that may be burned in certain European incinerators is in recognition of the fact that the continuous monitoring of actual incinerator emissions is still not possible, making it difficult to ensure that harmful emissions of combustion by-products and heavy metals will not occur.

Despite the fact that relatively little untreated hazardous waste is disposed of on the land in Europe, several European nations have continued to upgrade disposal techniques for treatment residues. This upgrading is being done so that the shift to treatment technologies can truly protect groundwater resources. Although landfill designs have recently been improved in the United States, the push for better residual disposal technologies should not stop. The concept of landfilling, where waste is placed beneath the surface near the water table, is seriously flawed and should be replaced with more secure above-ground designs.

The most significant question for America raised by a review of European technologies is whether a European-style system of centralized facilities using standard incineration and treatment technologies should be pursued as the means to shift beyond land disposal, or whether American ingenuity will produce a different and possibly more flexible technological response. Although European technologies are being transferred to the United States, there is already evidence that a uniquely American response is developing that builds on European experience without necessarily duplicating European facilities.

This response eschews the potential economic benefits of centralization in favor of the less tangible benefits of a decentralized response. Few integrated treatment facilities are being sited in the United States, but instead, many generators are managing waste on-site, and smaller, more specialized facilities, treating only a few types of waste, are springing up nearer the point of waste generation. Several hazardous waste management firms are also staking their futures on portable treatment units that bring the treatment process to the waste rather than the waste to the treatment process. Such treatment technologies diffuse rather than concentrate risk, encounter less public opposition, and give the generator more control over waste management to avoid future cleanup liabilities.

Some of the innovative technologies being put to use in the United States today, although not as versatile as standard treatment technologies like rotary kiln incineration, promise greater efficiency and safety in return. These include sophisticated thermal destruction technologies such as plasma arc destruction, infrared thermal processing, wet air oxidation, and supercritical water oxidation, that destroy organic waste more efficiently than conventional incineration and reduce air emissions of toxic combustion by-products. Recycling technologies for inorganic waste are also beginning to replace standard physical chemical treatment processes that produce large quantities of sludge requiring disposal.

In the final analysis, there is still a need to develop adequate hazardous waste treatment capacity in the United States for the waste still generated that will no longer be permitted to be disposed of on the land. The standard technologies that have been developed and proven in European hazardous waste treatment facilities may play at least a near-term role in the transition from land disposal. Their use should be temporary, however, since a new generation of innovative hazardous waste treatment technologies is being developed and put into use in the United States and the emphasis is shifting to waste reduction as the ultimate solution. Given these new developments, it may not be long before European engineers and policymakers begin traveling to the United States for lessons in hazardous waste management technology.

NOTES

1. Mogens Palmark, "The Danish System," paper presented at the 2nd International Symposium on Operating European Centralized Hazardous (Chemical) Waste Management Facilities, Odense, Denmark, September 1984.

2. Ibid.

3. Jens Kampmann, "Benefits from and Problems Experienced with the Danish Hazardous Waste Management System," paper presented at the 2nd International Symposium on Operating European Centralized Hazardous (Chemical) Waste Management Facilities, Odense, Denmark, September 1984.

4. Arne Kristensen, "Operating the Rotary Kiln Incinerators at Kommunekemi," paper presented at the 2nd International Symposium on Operating European Centralized Hazardous (Chemical) Waste Management Facilities, Odense, Denmark, September 1984.

5. Personal communication with Per Riemann, Plant Engineer, Kommunekemi, June 1983.

6. Palmark, "The Danish System."

7. Ibid.

8. Christina Sund, "Physico-Chemical Processing Options," paper presented at the 2nd International Symposium on Operating European Centralized Hazardous (Chemical Waste Management Facilities, Odense, Denmark, September 1984.

9. Ibid.

10. Kirsten Warnoe, "The Controlled Landfill," paper presented at the 1st International Symposium on Operating European Centralized Hazardous (Chemical) Waste Management Facilities, Odense, Denmark, September 1982.

11. Ibid.

12. Kampmann, "Benefits from and Problems Experienced with the Danish Hazardous Waste Management System."

13. Peter Løvgren, "The Danish System," paper presented at the 1st International Symposium on Operating European Centralized Hazardous (Chemical) Waste Management Facilities, Odense, Denmark, September 1982.

14. U.S. Environmental Protection Agency Science Advisory Board, "Report of the Committee on the Incineration of Liquid Hazardous Waste," Washington, D.C., 1985.

15. Thomas Petzinger, Jr. and Matt Moffett, *Wall Street Journal*, August 26, 1985, p. 1.

16. Kathryn E. Kelly, "Comparison of Metals Emissions Data from Hazardous Waste Incineration Facilities," paper presented at the 78th Annual Meeting of the Air Pollution Control Association, Detroit, Michigan, June 1985.

17. Personal communications with Dr. Ranier Meixlsperger, Bavarian State Agency for Environmental Protection, August 1983; Franz Defregger, Director, Waste Management Division, Bavarian State Ministry for Regional Development and Environmental Affairs, August 1985.

18. Personal communications with Peter Solyom, Swedish Water and Air Pollution Research Institute, August 1985; Hans Christoph Boppel, member of the Parliament of Hessen, Federal Republic of Germany, August 1985.

19. Heinz Womann, "Experiences with Industrial Waste Incineration at BASF," *Energie* 23, no. 11 (November 1971); H. W. Fabian and M. Schoen, "How Bayer Incinerates Wastes," *Hydrocarbon Processing* 58, no. 4 (April 1979): 183–92.

20. Womann, "Experiences with Industrial Waste Incineration."

21. Personal communication with Per Riemann, Plant Engineer, Kommunekemi, Nyborg, Denmark, June 1983; Arne Kristensen, "Rotary Kiln Incinerators for Organic Chemical Waste," paper presented at the 1st International Symposium on Operating European Centralized Hazardous (Chemical) Waste Management Facilities, Odense, Denmark, September 1982.

22. Personal communication with Carl Otto Zubiller, Director of Waste Management, Hessian Ministry for Labor, Environment, and Social Issues, Wiesbaden, Federal Republic of Germany, July 1985.

23. Personal communication with Dr. Gerhart Vogel, Technical University of Vienna, Vienna, Austria, August 1985.

24. Personal communication with Carl Otto Zubiller, Director of Waste Management, Hessian Ministry for Labor, Environment, and Social Issues, Wiesbaden, Federal Republic of Germany, August 1983.

25. Jens Moller, "Dry Scrubbing of Hazardous Waste Incinerator Flue Gas by Spray Dryer Absorption," paper presented at the 2nd International Symposium on Operating European Centralized Hazardous (Chemical) Waste Management Facilities, Odense, Denmark, September 1984.

26. Alois Scharsach, "The Regional Treatment Facility for Organic Hazardous Waste of the State of Hesse in Biebesheim, Germany," Von Roll Ltd., Zurich, Switzerland, 1983 (unpublished report).

27. William Viken and Per Waage, "Treatment of Hazardous Waste in Cement Kilns within a Decentralized Scheme: The Norwegian Experience," United Nations Environment Program. Industry and Environment Special Issue (1983), pp. 74–77.

28. "Destruction of Hazardous Waste-Emission of Organic Matter," summary of report concerning test burning of hazardous waste at Norcem Slemmestad plant for the Norwegian Ministry of Environment, Norcem Engineering, Oslo, Norway, June 1985.

29. Personal communication with Trygve Sverreson, Norcem, Oslo, Norway, July 19, 1985.

30. "Destruction of Hazardous Waste-Emission of Organic Matter."

31. Viken and Waage, "Treatment of Hazardous Waste in Cement Kilns within a Decentralized Scheme: the Norwegian Experience," pp. 74–77.

32. *Guide pour l'élimination et la valorisation des déchets industriels* [Guide for Disposal and Recycling of Industrial Waste], French Ministry of the Environment, 1984.

33. Robert Mouringham and Harry Freeman, "Hazardous Waste Incineration in Industrial Processes," in U.S. EPA, Hazardous Waste Engineering Research Laboratory, *Proceedings: International Conference on New Frontiers for Hazardous Waste Management*, EPA/600/9–85/025 (Cincinnati, Ohio, September 1985), p. 534.

34. For a relatively non-technical description of these technologies, see California Office of Appropriate Technology, *Alternatives to the Land Disposal of Hazardous Waste: An Assessment for California*, 1981; and U.S. Congress, Office of Technology Assessment, *Technologies and Management Strategies for Hazardous Waste Control*, Washington, D.C., 1983.

35. Norbert Amsoneit, "Waste Oil Separation at the Schwabach Facility," paper presented at the 2nd International Symposium on Operating European Centralized Hazardous (Chemical) Waste Management Facilities, Odense, Denmark, September 1984.

36. Sund, "Physico-Chemical Processing Options"; information about the ZVSMM facility was gained from a plant tour conducted by Dr. Norbert Amsoneit, Chief Chemist, ZVSMM, Schwabach, Federal Republic of Germany, July 1985.

37. Personal communication with D. LeMarchand, French National Agency for Recycling and Disposal of Wastes, Angers, France, June 1985.

38. Personal communications with H. Ruckel, Director, and Norbert Amsoneit, Chief Chemist, ZVSMM, Schwabach, Federal Republic of Germany, July 1985.

39. Ibid.

40. The 1984 amendments to the Resource Conservation and Recovery Act set out a program to restrict the land disposal of untreated hazardous waste and to set treatment standards. 42 U.S.C. § 6924.

41. Franz Defregger, "Status and Trends on the Management of Industrial Hazardous Waste in the Federal Republic of Germany," United Nations Environment Program Industry and Environment Special Issue (1983), pp. 19–20; personal communication with Dr. Ranier Meixlsperger, Bavarian State Agency for Environmental Protection, Munich, Federal Republic of Germany, August 1983.

42. Personal communication with Lars Llung, Manager, SAKAB, Norrtorp, Sweden, August 1985.

43. *Konzepte und Methoden der Sonderabfall beseitigung*, Protokoll des Wiesbadener Sonderabfall-Seminars, Die Grünen im Hessischen Landtag, June 23, 1984 [*Concepts and Methods for Special Waste Management*, Procedings of the Wiesbaden Special Waste Seminars, The Greens in the Hessian State Parliament, June 23, 1984], p. 3–5.

44. Kirk W. Brown, "Landfills of the Future: Above Ground and Above Board," in Bruce Piasecki, ed., *Beyond Dumping: New Strategies for Controlling Toxic Contamination*, (Westport, Conn.: Quorum Books, 1984), pp. 191–197.

45. Personal communication with J. van Zijst, Dutch Ministry of Housing, Land Use Planning, and Environment, Leidshendam, June 1985.

46. D. den Ouden, AVR-Chemie, "Hazardous Waste Management in the Netherlands," paper prepared for the International Solid Waste Association Working Group on Hazardous Waste, Hamburg, F.R.G., June 1985.

47. Gunnar Johnsson, "Underground Disposal at Herfa-Neurode," Kali and Salz AG, Kassel, Federal Republic of Germany, 1981.

4

Alternatives to Ocean Incineration in Europe _____

BRUCE PIASECKI AND HANS SUTTER

HISTORICAL BACKGROUND: WHY EUROPEANS BELIEVED OCEAN BURNING NECESSARY

Advocates for improved management of hazardous wastes in America often refer to Europe's ocean incineration efforts as yet another option for shifting the burden of waste management off the land. Yet for most Americans there is no single source of information and analysis of the European experience with ocean burning. This significant oversight is especially noteworthy considering the extent of European experience: While America has experimented with thirteen trial burns to date, the West Europeans have monitored the results of several hundred of such burns. What follows attempts to fill this gap by examining the West German response in detail. Since the West Germans were principal founder and primary contributors to at-sea incineration, their disillusionment with the approach serves as a telling gauge of European developments.

After an explanation of why Europeans first believed ocean burning necessary, West Germany's legal and regulatory efforts to phase out the technology are described. Faced with mounting control difficulties and select instances of contamination, the German government evolved a three-tier management strategy to secure alternatives to ocean burning. The policy logic and technical details of this successful effort are explored in the final sections of this essay.

During the 1960s the generation of chlorinated hydrocarbons increased rapidly with the expansion of the chemical industry. Much of this new waste was clearly regarded as hazardous, demanding special handling and management. Until the mid-1970s in Europe, the disposal of these chlorinated hydrocarbons created serious waste management problems. An appropriate disposal technology was conspicuously absent.

Examples of abusive practices of the past are commonplace. Thirty-eight thousand barrels containing chlorinated hydrocarbons were dumped directly into the North Sea between 1963 and 1969 alone.[1] In addition, the North Sea was used as a dumping ground for the bulk of chlorinated hydrocarbons (EDC-tar residues) generated by the popular and growing vinyl chloride production process. Remnants of these EDC-tar residues dot the North Sea like the PCB hot spots along the Hudson River.[2] Other unsound tactics used to get rid of chlorinated hydrocarbons included direct discharge into rivers or the sea, illegal dumping, landfilling, or improper combustion in incineration facilities lacking acid gas treatment of exhaust emissions. In such a setting, the market to support a capital-intensive treatment approach was simply missing.

Nonetheless, it was during the 1960s that some scientists proposed high-temperature burning as an efficient method for destroying chlorinated hydrocarbons. They recognized that an unfortunate side effect of the combustion process of these substances would be the emission of hydrochloric acid (HCL) and recommended equipping land-based incinerators with air pollution control devices to remove acidic gases and hazardous particles. Until the 1970s, however, the use of flue gas scrubbing was not widely practiced in Europe; thus the absence of emission control units on the first at-sea incinerators received little scrutiny.

When ocean burning started in Europe in 1969, it was used by two large West German chemical firms: Bayer AG of Leverkusen and Solvey & Cie of Rheinberg. These firms employed ocean vessels primarily for the burning of the chlorinated hydrocarbons generated in the production of chlorinated benzenes, vinyl chloride, and propylene oxide. The fostering concept behind ocean burning was that the main air pollutant emitted from the combustion of chlorinated materials, hydrogen chloride, would be absorbed by the seawater via droplets generated from the ocean's humidity. The buffering capacity of the seawater would, according to the notion that dilution was the solution, neutralize the acidic input.[3] Thus no costly scrubbers would be necessary. Cheaper disposal costs for the problematic chlorinated wastes were at last available for Europe.

Table 4.1 shows the quantities incinerated off the coast of Europe between 1969 and 1983. The amounts rose from 4,000 tons per year in 1969 to approximately 100,000 tons per year in 1979. Since 1982, however, partially due to mounting discontent with the performance of the equipment, there has been a consistent and significant decline in the quantities incinerated at sea. This decline should be noted by Americans as they debate the role of at-sea burning for hazardous waste treatment in the United States. Although the new Office of Technology Assessment (OTA) report issued to Congress in September 1986 makes note of this decline, the authors choose to ignore the forces of discontent behind Europe's recent decisions. Europeans, common wisdom claims, have simply placed too much environmental stress on the North Sea with their oil production and past disposal practices. Thus they intend to reduce further ocean

Table 4.1
Amounts of Waste Incinerated at the North Sea (in Tons)

1969	4,000	1977	55,400[1]
1970	8,000	1978	67,500[1]
1971	28,000	1979	107,000
1972	66,000	1980	108,000
1973	87,000	1981	100,600
1974	85,000	1982	97,526
1975	85,000	1983	80,234
1976	55,000		

[1]Only wastes loaded in Antwerp.

Source: H.W. Fabian, "Verbrennung chlorierter Kohlenwasserstoffe," *Umwelt*, January 1979, p. 12; Annual Reports on the Activities of the Oslo Commission.

burning. The question that remains is: Why are Americans prepared to pursue such a capital-intensive, environmentally unsound strategy when the founding nations of the approach now perceive it as obsolete? In an effort to achieve perspective on this question, the following account traces West Germany's concentrated efforts to phase out ocean burning.

Table 4.2 shows both the amount and origin of waste, as reported to the Oslo Commission for 1981 through 1983. By 1983 the total amount of waste delivered for incineration at sea was only 80,234 tons. Compared to 1981, this 20 percent reduction across all nations is telling. The Germans, moreover, reduced wastes from 58,561 tons in 1981 to 37,117 tons in 1983. Germany's reduced waste volume was a direct result of the nation's law calling for a strict adherence to the Oslo and London conventions, the two key international agreements seeking a phaseout to ocean burning. The significance of the German effort is further exemplified by the size of its declining contributions: in 1981 58 percent of the total waste sent to sea was German; by 1983 the reduction effort cut it to 46 percent of the total.

It is important to distinguish between the amount of waste delivered for ocean burning in a given year and the amount actually burned (Table 4.3). The discrepancy between these two figures demonstrates the difficulties in monitoring the approach. Sometimes the capacity of storage tanks in the ports of loading is so large that a delivered waste sits in storage until the next year. Sometimes wastes are transported down rivers in barges between the point of collection and the incineration vessel, making it hard to include the shipment in the conventional inventories.

Table 4.2
Origins of Wastes Delivered for Ocean Burning (in Tons)

COUNTRY OF ORIGIN OF WASTE	1981	1982	1983
AUSTRIA	126	512	171
BELGIUM	9,172	10,650	12,554
FRANCE	11,914	9,487	7,029
FEDERAL REPUBLIC OF GERMANY	58,561	39,560	37,117
FINLAND	—	2,750	—
IRELAND	40	—	—
ITALY	471	3,401	2,359
NETHERLANDS	7,483	17,970	4,058
NORWAY	3,356	4,392,	5,852
SPAIN	21	191	390
SWITZERLAND	3,653	3,679	2,735
SWEDEN	5,065	3,631	5,867
UNITED KINGDOM	811	1,303	2,102
TOTAL	100,673	97,526	80,234

Source: Chart constructed from official files of Dr. Hans Sutter, Umweltbundesamt, 1986.

In Table 4.3 the total amounts of waste collected for ocean burning are compared with the amounts collected in Antwerp. Since Antwerp is by far Europe's largest port for ocean burning, the authors believed it useful, in light of American concerns over the traffic and extent of stress at the waterfront facilities, to provide American readers with a sense of the scale of activities at Antwerp. Such a venture is massive in scope, highly centralized, and might prove quite difficult to site today considering America's political culture. Since OTA suggests that America could consume nearly 10 percent of its massive waste stream at sea, it is only accurate to translate these figures to a dozen Antwerp docks on America's primary coastlines.

Ocean burning was conceived to destroy only a special kind of chlorinated liquid waste; thus it is useful to keep in mind the usual means by which these special liquids are generated. Conventionally, these wastes occur in either very

Table 4.3
Amounts Delivered and Incinerated at Sea

	1981	1982	1983
Total amount delivered in port of loading for ocean burning (tons)	100,673	97,526	80,234
Total amount incinerated at sea (tons)	100,673	93,139	85,676
Amount incinerated from Antwerp (tons)	72,997	68,463	69,975

Source: Annual Reports on the Activities of the Oslo Commission.

small amounts (5–50 tons per month) with a low chlorine content from many different industrial processes, or in large amounts with a high chlorine content specifically from select processes within the chemical industry. In addition, the limitations associated with ship-based incinerators further restrict the range of eligibility. For example, not only must all the wastes entering the incinerator be liquid, but also none of the wastes may have a particulate content above 2 percent. The combined effect of sharp differences in generation and these eligibility restrictions is that waste for at-sea burning demands special and frequent sampling and testing in order to ensure efficient combustion.

The management task is further complicated by the fact that certain high-risk wastes fit within these restrictive parameters but should not go to sea. Certain chlorinated hydrocarbons such as polychlorinated biphenyls (PCBs), polychlorinated terphenyls (PCTs), and dioxin-containing wastes are strictly prohibited in Germany. This absolute prohibition is enforced throughout Germany because it has not yet been proven to the satisfaction of federal authorities that the necessary destruction efficiency can be reached at sea for these particular high-risk substances. The significance of these restrictions cannot be ignored any longer within the United States, where both dioxin and PCBs remain the most visible subjects of controversy. For many years the dioxin tragedy at Times Beach informed the resistance from the general public. Today it is also a fiercely controversial subject in the political debates. Nonetheless, both PCB- and dioxin-containing loads have been destroyed at sea within American waters.

The differences among the small and large generators in Europe and the United States lead to radically different storage strategies, further complicating the regulation and permitting requirements of at-sea ventures. Within Germany, the many small quantities are individually transported to storage tanks in Essen and Mannheim with a capacity of 1,000 tons and 1,500 tons.[4] From these two transfer sites, the wastes are shipped on the Rhine by barge to larger storage tanks at Antwerp, Belgium, where the waste is then loaded onto the incineration ship. In contrast, the wastes generated by major chemical firms are normally stored in private on-site facilities and then shipped directly to Antwerp. Depending on the available capacities of the incineration ships, the wastes can be loaded directly, but usually an extended storage period is required.

Because of the conflicting properties of the many different processes producing the small-quantity wastes, collection, transport, and storage of these smaller loads present considerable management challenges. Each waste producer, if responsible, must analyze the wastes for a number of possible complications. The chemical pretreatment of wastes, for instance, must be completed meticulously to prevent corrosion of storage tanks. Moreover, analysis of the small quantities often reveals wastes that must be excluded—after the expense of collection—because they do not meet the treatment requirements. These factors make the cost of ocean burning for the producer of small quantities more than twice that of most large producers.

It is this class of collection, transport, and storage difficulties that feeds much of the resistence in the United States. In May 1986 EPA official Patrick M. Tobin in his hearing officer's report to the EPA administrator summarized the Philadelphia hearing as focusing its critique on these infrastructure problems. According to Tobin, the people of Philadelphia were more distressed by the associated risks of collection, transport, and storage than the actual efficiency or design of the incinerators. They explicitly doubted the efficacy of waste transport by barge down the Delaware; moreover, they questioned the viability of a loading facility in the fragile ecosystem of the coast.

The performance history of Europe's incineration ships is not reassuring. Table 4.4 shows that the record is spotty and somewhat unpredictable. In May 1983 the license for the *Matthias II* was cancelled by German authorities (German Hydrographic Institute with the Federal Environmental Agency) when dioxins were discovered in the converted tankship's exhaust gases. The *Matthias III* operated only for a short time, since the market failed to provide the ship with enough waste for it to operate economically. The demise of these two ships, the latter specially built for ocean burning, suggests that such highly centralized systems demand a greater uniformity and regularity to the waste market than normally exists in the United States. Of the three ocean-going vessels still in operation *(Vulcanus I, Vulcanus II,* and *Vesta),* all are Type II double-hull tankers approved to carry chemicals by the International Maritime Organization. The age of this fleet is a concern: *Vulcanus I* was built in 1965 as a tankship and

Table 4.4
Incineration Ships

NAME	MATTHIAS I	MATTHIAS* II	MATTHIAS III	VULCANUS I	VULCANUS II	VESTA
In service	1968	1970	1975/6	1972(83)	1982/3	1979
Out of service	1976	1983	1977	—	—	—
Number of ovens	1	1	1	2	3	1
Waste destruction capacity ton/hr	3–6	10–12	20–40	20–25	max. 35	10
Waste loading cap. metric tons	550	1.200	1.500	3.500	3.300	1.400

*Decommissioned because of dioxin contamination.

Sources: L. Barniske, "Technische Gesichtspunkte bei der Verbrennung von Chlorkohlenwasserstoff-Abfallen," *Abfallbeseitigung auf See, Beihefte zu Müll und Abfall*, (1985), pp. 47–56; Länderarbeitsgemeinschaft Abfall, "Vergleichende Zusammenstellung über die Verbrennung chlorierter Kohlenwasserstoffe an Land und auf Höher See," government report, 1983.

rebuilt as an incineration ship in 1972. *Vesta* (1971) and *Vulcanus II* (1982) were specifically designed for ocean burning by their owners.

The European experience with these ships has led to significant changes in site designation over the years. Until 1978 the ships operated in an area only 10–15 miles from the coast of the Netherlands. But complaints from the coastal people about exhaust gases from the ships forced the vessels to their more remote present location, where the closest coastal areas are now over 70 nautical miles away.[5] Citizens on the Dutch shore once could see the flames from their homes and often spoke of the smells of the acidic emissions. It now takes about 24 hours to reach the incineration area from Antwerp. With a radius of 15 nautical miles, the current North Sea site is a well-defined circular area where the water's depth exceeds 40 meters. In bad weather an auxiliary site closer to the coast is occasionally still used. But again, the need for these more distant sites was answered only after initial sites proved insufficient.

When first adopted in Europe, ocean burning was a progressive technology preferred over ocean dumping and landfilling. Today, however, it is no longer perceived as the final solution to the problem of chlorinated wastes. Besides its history of operating and maintenance problems, ocean burning cannot recover any of the waste's valuable components for subsequent recycling. This flaw is fundamental. In response, technologies have been developed that provide generators with a satisfactory and lasting land-based option to these waste burdens. In the past, however, the ease of ocean burning severely blocked the introduction of better solutions and kept a generation of more reliable on-site treatment systems on hold. In the next two sections Germany's efforts to facilitate the use of these safer alternatives are discussed.

THE GERMAN LEGAL AND REGULATORY EFFORT TO CONTROL OCEAN BURNING

The Oslo and London conventions, the chief international agreements limiting ocean dumping and incineration, stimulated legislation in the Federal Republic of Germany to prevent marine pollution by waste disposal. The federal law that came into effect in December 1977 subjects all dumping and ocean burning ventures to administrative controls if the transaction starts from a German port or if the ship is flying a German flag.[6] Thus before 1978 the burning of wastes at sea occurred without the need of a permit.

The narrowed conditions of eligibility for an ocean-going permit indicate the disfavor with which German officials came to perceive an approach they first fathered. Article 2 of the 1977 law states that a permit can be granted if two prerequisites are fulfilled: first, since land-based treatment methods are explicitly preferred, a permit can be granted only when it is impossible to dispose of the waste on land without incurring public harm or disproportionate costs; second, the authorities can grant a permit only if "no adverse impacts on human health, living resources, or other uses of the sea" are incurred. In exceptional cases, it is possible to grant a permit without fulfilling both of the above prerequisites if the Federal Minister of Transport, the Federal Minister of the Interior, and the Federal Minister of Economics agree that the permit is urgently required on behalf of the public—as in an emergency situation. Moreover, this triple approval can be argued as answering the public interest only on a case-by-case basis.

This German law goes beyond the guidelines of the Oslo and London conventions on a number of points. Under the conventions' explicit mandate that priority be given to the land-based treatment of wastes, any German request for ocean burning is denied if a land-based alternative is available. But the determination of whether the wastes can be treated on land is not restricted to the traditional waste disposal practices: Germany has pushed the conventions further by showing that land-based treatment options can also include changes in waste-generating processes to reduce waste production as well as reuse and recycling. This means that once it has been established that the wastes may be treated in any innovative fashion on land, the possibility of ocean burning will no longer be considered by the German authorities.

The German law also goes beyond the requirements of the Oslo and London conventions in interpreting what constitutes a "concern." Any adverse impact at sea is already regarded as a justified concern if past experience or scientific studies show that the adverse impact is probable. Thus the closing of the *Matthias II* had multiple effects, shifting the burden of proof away from the government and onto the generators. The Germans have already used the "justified concern" provision to change the ocean dumping of certain acid wastes. Moreover, it is sufficient for experts to deem adverse effects at sea possible solely on the basis of scientific findings, without the need to furnish exact operational proof of malfunction or release. Thus both past, single-episode experiences and new scientific results qualify with equal force as "concerns" sufficient to deny at-sea requests.

In order to balance the judgment on what wastes may be treated on land, the Germans established an administrative procedure that involves experts from a wide range of federal and state agencies.[7] This procedure, which strategically depoliticizes the judgment by rerouting the weight of the decision away from Bonn and into the technical research institutes, requires that the applicant for at-sea burning submit the request to the German Hydrographic Institute (Deutsches Hydrographisches Institute or DHI), a technical branch of the Ministry of Transport and the authority responsible for issuing any permits for disposal at sea. The DHI then consults with the Federal Environmental Agency (Umweltbundesamt, or UBA) to determine if a treatment on land is possible. The UBA in turn consults various regional bodies and industrial firms and conveys its opinion to the DHI. In the end, the UBA is the authority responsible for the decision on whether land-based treatment options are available or not. If the UBA identifies a land-based alternative, the DHI must reject the approval, and no permit for incineration at sea can be granted. If the decision on alternatives is negative, the DHI must study the possible consequences of ocean burning on the marine environment. By having DHI's denial hinge on UBA's assessment of land-based alternatives, the DHI is in a sense freed from direct applicant pressure.

Should UBA's decision state that no land-based alternative is available, the mechanism of further approval involves ten federal and regional services. A sample of the services consulted usually would include health departments, the Biological Institute of Helgoland, and the Federal Agency for Management of Waters and Navigation. Again, the strategy is to give the DHI the authority, but to spread its decision across numerous technical branches of government. If all these hurdles are passed, the Germans discourage ocean burning yet again. The incineration permit is valid for only two years. This shorter period stands in stark contrast to the American deliberations over a ten-year operational permit.

With its firm priority for land-based treatment, the strength of the German program rests in the UBA's readiness to investigate and develop land-based opportunities. Each incineration permit applicant receives a thorough investigation by the UBA. Each waste stream greater than 600 tons per year must be described in full by the applicants. This baseline information contains data on the source of the waste, the nature of the production process that generates it, and the physical and chemical properties of the waste stream. Moreover, any research done by the applicant concerning possible waste reduction and recycling measures must also be reported as a part of the request. Based on this data, the UBA then performs its own research into land-based treatment alternatives. Such investigations include detailed inquiries and meetings with the applicants about the quality of the information submitted as well as sustained discussions with other industrial firms and scientific experts about the prospects of alternatives to at-sea treatment for the specific waste streams.

UBA's experiences with developing these alternatives for the troubling waste streams reveal an unexpected trend: only a minor part of the wastes requested for ocean burning are suited for destruction in land-based hazardous waste incinerators. Instead, the most promising options UBA has developed are based primarily on waste reduction and reutilization. The importance of this experience can no longer be ignored in the United States: without strong government sponsorship of reduction programs for highly chlorinated wastes, the pressure for ocean burning will only increase in America.

UBA's clear departure from traditional "end-of-the-pipe" thinking about these problematic wastes is noteworthy for a second reason. Not only does it demonstrate land-based alternatives via direct research programs, but it also sets rigid schedules by which applicants must change their production technologies to lessen their burdens on the sea. Often, for example, the UBA signals the need for a new on-site approach by stating as a condition of the incineration permit that after a certain date no further permit will be given to incinerate the same wastes at sea. This most explicit form of technological prompting has gone far in making German firms producing highly chlorinated wastes look hard at on-site modifications.

UBA's recommendations, moreover, are not completely at the mercy of costs. Often its land-based reduction alternatives are cheaper when measured across a few years. In those cases in which UBA's preferred alternative remains more expensive than ocean burning, these higher costs are not looked at as unjustified. Instead, in at least three recent cases incineration permits have been denied by the DHI when slightly more expensive alternatives were proven technically available by the UBA. Again, the primacy of this best available technology approach above costs would go far in the United States toward installing a strong new generation of waste reduction equipment on-site.

UBA's emphasis has, of course, met with considerable resistance from highly chlorinated waste producing segments of the German chemical industry. For them, as in the United States, incineration promises an easy, low-liability option for waste management without opening the can of worms called process change. As an end-of-the-pipe technology, at-sea burning needs little attention from production staff. It neglects the links between generator and handler as it widens the gap between production design and waste treatment. UBA's land-based alternatives, on the other hand, usually involve a restructuring of production techniques and at times even demand cooperation between different firms to recycle chlorinated hydrocarbons. How has the UBA been able to win such significant battles?

Experience has shown that the support for UBA's claims comes from segments of industry less sensitive than the firms directly generating the wastes in question. For example, when it is proven by UBA that certain German plants have the excess capacity to reprocess another firm's waste as raw materials, the targeted plant often expresses a willingness to accept the waste but the producing firm remains unwilling to deliver it. This common hurdle can be overcome in Germany because DHI has the authority to deny the producing firm's request for ocean burning as soon as the recipient firm expresses its willingness to serve as a land-based alternative. Left without its at-sea option, the unwilling generator often complies. A major justification for such arm-twisting rests in the UBA's reconceptualization of select high-risk wastes as valued resources. Once valued by production staff, these by-products are more vigilantly managed—and the stigma accorded waste is removed in the act of salvaging it.

CONTROL DIFFICULTIES AND ENVIRONMENTAL IMPACTS: A CHALLENGE TO MANAGEMENT

In addition to permit requirements, German incineration ships must be licensed for waste destruction.[8] In accord with international agreements, these licenses are granted only when an incineration firm demonstrates combustion and destruction efficiencies exceeding 99.9 percent.[9] Each license is valid for only two years and must be approved again by all the German authorities before the renewed applicant can burn.[10]

The scrupulousness of these renewal requirements played a significant role in closing down the *Matthias II*. The ship had already been operating for some time before these requirements came into place. In its initial licensing survey, the *Matthias II* had reached a destruction efficiency of 99.95 percent. Later renewal tests proved, however, that the initial test conditions did not represent normal operating conditions. The flow rate under the initial test was between 6 and 8 tons per hour (tph), compared with a normal commercial feed rate of 10–13 tph and a maximum rate of 15 tph. The wastes had been incinerated in the initial tests at temperatures of 1500 degrees centigrade compared with normal operating temperatures of 1350–1400 degrees centigrade. This discrepancy proved costly to the *Matthias II*'s owners. Further tests conducted under normal operating conditions resulted in the withdrawal of the vessel's license by the DHI in May 1983. The study concluded that chlorinated dioxins were found in the stack gases at concentrations that could adversely affect the marine environment.

According to the second set of licensing investigations, TCDDs in concentrations of 0.2 to 22.9 $\mu g/m^3$ with a mean value of 8.5 $\mu g/m^3$ were found in the exhaust gases. On the basis of these values, it is estimated that 40–50 grams of TCDDs entered the atmosphere per incineration voyage (computations were based on 100 hours' duration. These surprisingly high concentrations of TCDDs are environmentally hazardous both because of their great toxicity and because they are accumulated not only by mammals but also by fish. Moreover, 2,3,7,8–TCDD (dioxin), the most widely investigated and dangerous isomer, was inferred to be present also in the exhaust gas of the *Matthias II* on previous burns. After the German authorities revoked the license, Dutch authorities refused an application for *Matthias II* on the same grounds. Within weeks of the German report, the *Matthias II* was permanently decommissioned in Europe.

It is possible that the specific design of the incinerator on board the *Matthias II* was the cause of the high TCDD concentrations. The other ships (*Vesta, Vulcanus I* and *II*) are equipped with somewhat different designs, yet even these operating ships employ the same type of rotary cup burners. However, scientific information about the influence of incinerator design upon the formation of dioxins is not sufficient to draw a clear conclusion in this respect.

This element of uncertainty must be considered carefully, since decisions based only on destruction efficiency or design characteristics are innately suspect. Destruction efficiency provides an inadequate measurement of incineration emissions, as the case of the *Matthias II* dramatically confirms. Indeed, the U.S. Environmental Protection Agency's Science Advisory Board reports that destruction efficiency was never meant to fulfill an emissions control function, but instead was intended only as a policy guideline for evaluating achievable standards of incineration. Destruction efficiencies measure only the proportions, not the amounts or kinds, of toxic chemicals emitted. Since such measurements do not provide any limit or regulatory ceiling in themselves, many Americans believe that regulations based on destruction efficiency alone cannot properly ensure human health.

Similar doubts have been voiced against evaluations based solely on design characteristics. Edward W. Kleppinger, an environmental consultant, and Desmond H. Bond, a chemical engineer, say of the Vulcanus incinerator that its stacks cannot be sampled adequately to determine gas flow, gas composition, or particulate content. Thus they claim both the Vulcanus and Apollo designs unverifiable.[11]

With similar uncertainties in the wind, German officials were especially earnest in testing the *Vesta* when it came into service in 1979. As a ship specially designed for at-sea burning, it's newness provided the UBA with a clean slate. Thus the emission concentrations of carbon monoxide, carbon dioxide, oxygen, and nitrogen oxide were all recorded by the UBA, and the total quantities of organic hydrocarbon and inorganic gaseous chlorine compounds were measured at frequent intervals.[12] Gas chromatography identified the major constituents in the waste inputs; these were compared to the exhaust gas constituents after combustion. The tests revealed a combustion efficiency of 99.5 percent ± 0.05 percent. Moreover, the destruction performance averaged over the incineration period of the respective waste exceeded 99.9 percent. To the surprise of some, these investigations found that the *Vesta* complied with all the requirements laid down in the international conventions. Thus the German authorities granted approval.

The *Vulcanus I* and *II* do not fly the German flag. They are, instead, licensed and controlled by the Dutch authorities. The results of Dutch investigations suggest that these vessels also comply with the international rules on incineration at sea. During the initial surveys of the incinerator of the *Vulcanus II*, for example, the destruction efficiency was approximately 99.995 percent according to Dutch officials.[13]

Another challenge ocean vessels present involves monitoring after the licensing test burns. German officials keep tabs on the ships in two ways: through installed measuring instruments and through films taken by inspection planes. Since the films are reviewed only on a monthly basis, the monitoring equipment provides the closer survey. UBA's equipment records numerous parameters; the most significant include wall temperatures of the incinerators, feed rates and amounts of waste pumped to the incinerator, concentrations of carbon dioxide, carbon monoxide, and oxygen in the combustion gases, the dates and hours of the operation, the status of the waste pumps and flame sensors, and the vessel's call sign and position. In addition, the at-sea firms report on the identity of their waste suppliers, the types and amounts of wastes delivered to the storage tanks and loaded on board, and the weather conditions during the periods of incineration. A number of design specifications also are required to minimize risks. Permit requirements prohibit ships from carrying their own pumping equipment for the loading of the tanks; they must instead employ the fixed dockside pumps. Storage tanks, moreover, are equipped with level indicators and overfilling safety devices to prevent accidental waste spillage. An automatic feed shut-off stops the flow of waste if the temperature drops below a designated level.

Despite these precautions, new data regarding the formation of dioxins through the combustion of chlorinated hydrocarbons suggests more negative environmental implications for ocean burning. Recent research in Sweden indicates that highly toxic chlorinated organic compounds can be sythesized if the organic compounds in the waste are not completely destroyed during incineration. That these chlorinated dioxins and furans have not been detected during test burns, except in the case of the *Matthias II*, further intensifies the uncertainties surrounding ocean burning.

Researchers differ in their opinion on whether this is simply a result of the sampling and analytical procedures employed. Some argue that accurate, valid sampling of the exhaust gases has not been conducted.[14] Conventionally, firms acquire their emissions data from a fixed probe location. Instead of this method, some scientists contend, a sampling procedure should be established that obtains a gas sample across many points within the stack. The concern is that if gas flow in the stack varies—and it does, considerably, in normal operations—condensed materials and particulates may be concentrated in unsampled areas within the stack. In addition, a more rigorous method for collecting and evaluating stack particulates is urgently needed. This necessity is heightened by the fact that the small quantities of waste collected for burning often contain up to 2 percent particulate matter. With these basic research questions still unanswered in Europe, America's research program gains in its international importance.

STRATEGIES FOR TERMINATING OCEAN BURNING

Many Americans assume that if we do not burn at sea, our only alternative is land-based hazardous waste incineration plants. The German experience, however, demonstrates that this assumption is incorrect. Land-based incinerators have played a minor role in Germany's strategy to terminate ocean burning. Instead, the majority of Germany's wastes once sent to sea have been managed by recycling, reuse, or waste reduction techniques. This suggests that ocean burning is not only publicly unacceptable to the Germans but also technically unnecessary to the degree anticipated by some American policymakers who are currently scrutinizing policy alternatives for the United States. This section reconstructs the means by which Germany reordered its priorities in favor of alternatives to ocean incineration.

The strength of the German approach lies in its decision to match the different chemical qualities of its waste with distinct control strategies.[15] Rather than simply commingling all problematic chlorinated hydrocarbons in a single dockside stockpile, the Germans have established a three-tier management scheme based on the organic chlorine content of the wastes. Thus the chemical nature of the waste itself, as opposed to a managerial motive to achieve bulk quantities, shapes the German effort.

The logic of this divide and conquer approach is easily summarized. Wastes with a high chlorine content are difficult to destroy in incinerators. They require extensive flue gas cleaning, yet even then the chlorine content eventually migrates into waterways from the salts in the wet flue gas scrubbers or into the land from the residues of dry scrubbers. But the Germans found that these highly chlorinated wastes have a considerable potential for reuse as raw materials. In this context, their high chlorine content is an asset to various chemical processes.

The opposite is true of wastes having low chlorine content: they are not easily reused because they lack raw material value. Yet they can be destroyed in hazardous waste incinerators since they do not create serious residue problems

Table 4.5
Amounts of Waste Applications for Ocean Burning (1981)

Waste	Chlorine Content (% Weight)	Amount (tons per year)	Chlorine load	Source
Highly chlorinated	40 - 85, 65 (average)	60.000	39.000	Chemical industry
Moderately chlorinated	10 - 40, 30 (average)	40.000	12.000	Chemical industry, other industries
Slightly chlorinated	up to 10, 5 (average)	20.000	1.000	Chemical and other industries

Source: H. Sutter, "Tendenzen bei der Verbrennung von chlorierten Kohlenwasserstoffen auf
    See," *Müll und Abfall*, April 1984.

or demand extensive flue gas cleaning. Thus, as the chlorine content increases, waste reutilization becomes increasingly practical whereas destruction becomes increasingly difficult. With this basic tenet in mind, the Germans have built their three-tier response to the problem.

Table 4.5 summarizes Germany's segregation of the waste stream into three distinct categories; table 4.6 summarizes the distinct methods and exemplary processes used for the different waste types. Table 4.5 also provides a summary of the quantities of the three categories sent to sea in 1981, when two German incineration ships were destroying most of the waste sent to sea. Although this situation has changed with only one ship now flying the German flag, the figures remain telling since the processes proved functional in Table 4.6 are appropriate for wastes of any nation at any time.

About 90,000 of the total 120,000 tons reported in Table 4.5 were produced by German firms. The rest came primarily from France, Belgium, the Netherlands, Sweden, and Norway. Half of the total amount (60,000 tons) was highly chlorinated waste. It is important to note that a single industry, the chemical industry, created this category of waste ripe for reuse. Moderately chlorinated wastes (40,000 tons) were produced by both the chemical industry and other industrial sectors, where they were most often generated by the use of chlorinated

Table 4.6
Land-Based Options

Waste	Method	Product	Process (examples)
Highly chlorinated	Chlorination Oxichlorination Perchlorination	Trichloroethylene Trichloroethylene & Perchloroethylene Perchloroethylene & Carbon tetrachloride	Chloe PPG CWH, Hoechst
	Combustion	Hydrogen Chloride Hydrochloric acid Heat	Uhde, BASF, Chloe, Nittetu, John Zink Stauffer
Moderately chlorinated	Combustion	Hydrochloric acid Heat	similar to highly chlorinated waste generators
	Distillation Extraction	Solvents	many methods
Slightly chlorinated	Hazardous waste incineration	HCl (not recovered) Heat	HIM, GSB

Source: H. Fricke, "Zwangsanfall von Ruchstanden und ihre Bereitigungsmöglichkeiten an Land," *Abfallbeseitigung auf See, Beihefte zu Müll und Abfall* 17 (1980).

solvents in metal finishing or other cleaning processes. The smallest amount is made up of slightly chlorinated wastes (20,000 tons) originating from the same sources as the moderately chlorinated wastes.

The basic concept empowering the German strategy, then, is that the different waste categories must be directed to the appropriate land-based options (Table 4.6). This implies that highly chlorinated residues should be used as raw materials in the chemical industry; moderately chlorinated wastes also have a raw material value and should be channelized into reuse; and slightly chlorinated wastes with no material value should be destroyed in technically sophisticated hazardous waste incinerators. This is what the Germans actually require. What follows looks more closely at the technical solutions employed for each of the segregated waste qualities.

Reusing Highly Chlorinated Wastes: The Benefits of Government Involvement

Only a handful of chemical processes produce the bulk of the highly chlorinated wastes in question (ethylene dichloride [EDC], vinyl chloride [VC], propylene oxide, chlorinated methanes, perchloroethylene, and carbon tetrachloride). If contrasted with the mixtures of low-chlorine wastes, these highly chlorinated wastes are relatively pure, enabling their reutilization in a variety of processes. Table 4.7 displays the principal chemical processes that generate chlorinated organic residues along with their respective chlorine content.

Chlorinated organic chemicals are manufactured in great volumes, with two to ten thousand tons generated per year at each of the typical source points. Thus source-related reutilization methods are the most sensible, since they cut handling and transportation problems to a minimum. In the case of moderately chlorinated wastes, however, especially those coming from metal-cleaning processes, one does confront serious managerial problems.

Highly chlorinated hydrocarbon residues have a long history of reuse within the German chemical industry. The most prevalent methods are the direct use of high-quality chlorinated hydrocarbon residues in perchlorination processes for the production of tri- and perchloroethylene and the utilization of chlorinated hydrocarbon residues for the production of hydrogen chloride and hydrochloric acid. Table 4.8 displays how highly chlorinated hydrocarbon residues from the EDC/VC-production process, for instance, were reused in Germany in 1981. From a total amount of 22,600 tons, nearly 55 percent was used as raw material in perchlorination processes, with another 15 percent used as raw material to produce hydrogen chloride. The remaining 30 percent was burned at sea.

The benefits of this approach on-site are multiple. In addition to conserving chlorine, these residues serve as substitutes for valuable raw materials like acetylene and ethylene. The most important residues used as substitutes are the light ends from the EDC/VC-process and the 1,2–dichloropropane residues generated in the production of propylene oxide.

In the recent past, chemical firms applied to incinerate these residues at sea. When UBA identified that excess capacity existed in the perchlorination plants of other chemical firms, the waste-producing firms often proved reluctant to accept this alternative even if it was cheaper than ocean burning. Eventually, as it became clear that DHI would not grant any further permits for these wastes, the waste producer was forced to accept the land-based alternative.

For those highly chlorinated wastes that cannot be reused in perchlorination processes (either because of chemical qualities or market limits), other land-based options exist in Germany to obtain hydrogen chloride or hydrochloric acid.

Table 4.7
Principal Chemical Processes Generating Chlorinated Residues

Chemical Process	Specification of the Residues	
	Name	Chlorine content % Weight
EDC from chlorination	Light ends	60
VCM from EDC-cracking	Heavy ends	70
EDC from oxichlorination	Light ends	60
	Heavy ends	70
Propylene oxide by the chlorohydrin route	1.2-Dichloropropane	60
Chlorinated methanes	Heavy ends	70 – 85
Per and tetra from perchlorination	Heavy ends	80
Production of chlorinated paraffins	Chlorinated paraffins	10 – 50
Production of chlorinated benzenes	Heavy ends	10 – 50
Production of chlorinated agrichemicals	—	10 – 80

Source: H. Sutter, ''Tendenzen bei der Verbrennung von chlorierten Kohlenwasserstoffen auf See,'' *Müll und Abfall*, April 1984.

The three main routes for reuse of these by-products are extraction of a 30 percent concentration of hydrochloric acid; extraction of gaseous hydrogen chloride; and the direct introduction of flue gas into chemical reactors. What follows is a brief account of how each of these approaches are currently employed by a number of German firms to help phase out ocean burning.

Figure 4.1 shows the relationship between the cost of ocean burning and the value of the product (100 percent HCl) for each of these procedures. The basis for the calculation is a small-plant capacity of 4,000 tons per year and a waste with a chlorine content of 60 percent. The cost will decrease for plants of larger capacity because of the economics of scale inherent in capital costs. The cost of the incineration (no product value) is shown as the ordinate value (procedure

Table 4.8
Treatment of Residues from the EDC/VC-Production Process (1981)

Method of Treatment	Quantity (t)	%
Use in perchlorination	12,400	54.9
Incineration with HCl-extraction	3,400	15.0
Ocean burning	6,800	30.1
TOTAL	22,600	100.0

Source: H. Sutter, "Tendenzen bei der Verbrennung von chlorierten Kohlenwasserstoffen auf
   See," *Müll und Abfall*, April 1984.

1 costs 308 DM/t waste, procedure 2 costs 406 DM/t, and procedure 3 costs 360 DM/t). The product values that are necessary to make the incineration costs zero are shown at the abscissa. Thus procedure 1 needs a product value of 395 DM/t 100 percent HCl, procedure 2 needs 520 DM/t, and procedure 3 needs 480 DM/t.

The connecting lines between points depict the incineration costs as a function of the different possible product values. Figure 4.1 also displays the costs of ocean burning (220–250 DM/t) in reference to the product value that could be realized in recent years for a high-quality hydrochloric acid (about 320 DM/t 100 percent HCl). This comparison reveals that once we consider lost product value, ocean burning is significantly more expensive than reutilization for highly chlorinated wastes.

In addition to such cost-saving techniques as waste reuse at perchlorination plants, German firms recover the value of their highly chlorinated wastes by coupling on-site combustion with chemical processes that use the resulting hydrochloric acid (Figure 4.2). In fact, the German branch of Dow Chemical designed its Stade plant on this principle. Three different production units at Stade are equipped with an interconnected incinerator that produces hydrochloric acid from the chlorinated wastes. This acid is then reused in connected production processes. In total, these three Stade incinerators produce from 600 to 5,000 tons of reusable residues per year for Dow.

Although Dow opted for this land-based treatment well before most other firms in Germany, others are now rapidly eliminating their at-sea activities. BASF, for instance, phased out its large ocean burning efforts in 1981. All the wastes resulting from EDC/VC production in both BASF's German and Belgian plants now undergo land-based treatment. Using the procedure outlined in Figure 4.3, their incineration facility in Antwerp recycles between 5,000 and 6,000 tons of waste per year with a chlorine content greater than 68 percent. The plant not

Figure 4.1
Cost of Residue Incineration with HCl Recovery

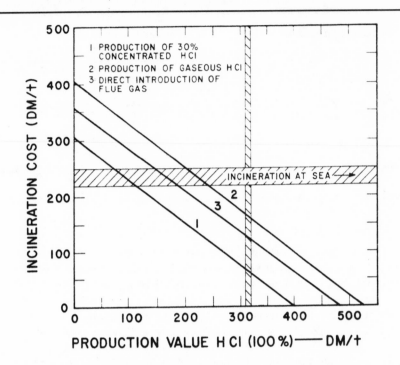

Source: H. Fricke, ''Zwangsanfall von Ruckstanden und ihre Beseitigungsmöglichkeiten an Land,'' *Abfallbeseitigung auf See, Beihefte zu Müll und Abfall* 17 (1980).

only produces recoverable heat, but it directly reuses a waste product in its oxychlorination process to produce EDC. Since BASF's treatment costs are similar to the costs of at-sea incineration, other firms are taking related actions.

Both the Chemische Werke Huls in Marl and the Wacker Chemie plant in Burghausen produce hydrochloric acid for reuse in other chemical processes. The Dutch firm Akzo Zout Chemie incinerates wastes with a 45–85 percent chlorine content into gaseous hydrogen chloride, which is then used in their oxychlorination process. This facility incinerates about 28,000 tons of waste per year. Presently, the plant incinerates waste from Akzo and a Swedish firm. Both of these waste streams were originally burned at sea. Other German firms reuse the hydrogen chloride in hydrochlorination reactions for chlorinated hydrocarbon production. Still others use it for the oxychlorination of vinyl chloride in diverse manufacturing processes.

The lesson from these activities is obvious, but often overlooked. A nation can induce a tremendous reduction in the amount of highly chlorinated wastes it sends to be burned at sea. Germany achieved this task mainly by requiring

Figure 4.2
Direct Introduction of Flue Gas

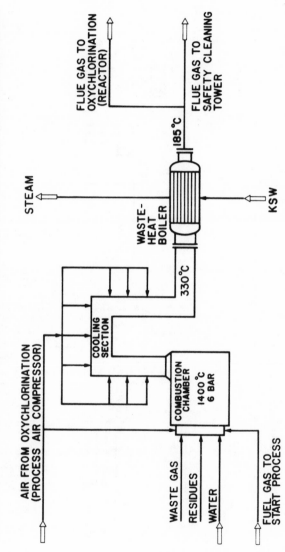

Figure 4.3
Production of 30%-Concentrated HCl

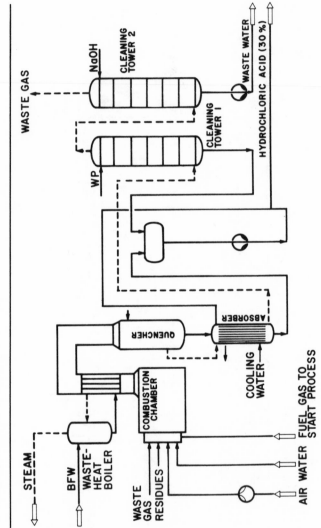

waste reuse in perchlorination processes. The effort was further enhanced by installing incinerators designed to produce hydrochloric acid or hydrogen chloride for reuse in closely coupled production processes. These land-based on-site reforms have proven so successful that by 1988 only one German firm will still send its highly chlorinated wastes to sea.

Moderately Chlorinated Wastes: Rerouted and Reused

The moderately chlorinated wastes are generated in the chemical industry as well as other industrial sectors (see Table 4.5). Within the chemical industry they are produced in large amounts (2,000–4,000 tons per year) as by-products of processes like the chlorination of benzenes. Another common source derives from the use of solvents in pharmaceuticals and agrichemical manufacturing. These wastes often contain not only halogenated and non-halogenated solvents but also a wide range of other impurities. Outside the chemical industry the use of chlorinated solvents in such cleaning processes as metal cleaning or dry cleaning also results in moderately chlorinated wastes, but these amounts rarely exceed 600 tons annually at each point source. Thus both the diversity of sources and the range of impurities found within the chlorinated wastes present a considerable treatment challenge.

In the past this range of wastes from both within and without the chemical industry was collected by waste-handling firms and transported to storage facilities in Mannheim and Essen. In this context, at-sea incineration appeared convenient, since the distinctions of source and impurity were never made. Instead, once large enough amounts were accumulated, the wastes from both facilities were sent to the port of Antwerp, marked for ocean burning.

Yet the German authorities were not content with this deceptively convenient system. They decided to confront the complexities of the waste streams' content directly. The federal government sponsored an investigation into the possibilities of utilizing the wastes collected in these storage facilities.[16] The explicit aim of the program was to develop land-based options for the moderately chlorinated wastes so as to further reduce ocean burning. Each waste shipment destined for ocean burning was analyzed for chemical and physical composition at the point of entry to the storage facilities. The results were classified before the shipments were placed in bulk storage.

Interestingly enough, recycling possibilities could not be secured for many single shipments, but the mixtures in the tanks showed promise for reuse. The Germans then developed a procedure to separate the mixtures into a water-based segment and an organic segment with a chlorine content of about 30–35 percent. Researchers within the UBA recommended that the two segments be treated separately: some of the water-based segment is now purified by extraction, and much of the organic segment is incinerated in a plant similar to the German plants used to incinerate highly chlorinated wastes (Figure 4.2). Heat is recovered in these incinerators and the resulting hydrochloric acid is mainly reused in metal treatment. Once again, acting as research initiator and intervenor, the government reduced the load formerly burned at sea and implemented a viable treatment method instead.

In light of these changes, the UBA has recommended the construction of a larger incineration plant.[17] Although detailed designs are still in progress, the Germans are planning to include the incineration of PCBs in this design. With the gradual replacement of PCB-filled electrical equipment occurring throughout Germany, the idea of tackling both the PCB problem and the moderately chlorinated waste stream in the same system gained strong support. Test runs with PCBs in a large-scale plant have been completed successfully and have shown that high-temperature, land-based incineration (Figure 4.2) can destroy PCBs effectively and produce hydrochloric acid. The design now looks for a yearly capacity of 20,000 tons of moderately chlorinated waste and 2,000 tons of PCB-containing liquids from transformers and condensers. If sited, this new facility will further divert the moderately chlorinated wastes presently sent to sea and enlarge the capacity to destroy PCBs. The final decision to install the large-scale plant has not yet been made, so until that date DHI still allows some moderately chlorinated wastes to be incinerated at sea.

In addition to the above rerouting of stored wastes, other moderately chlorinated wastes have been reduced on-site. This was achieved mainly by reworking chemical processes to minimize waste generation and by increasing the use of solvent recovery techniques. Although the latter development was clearly enhanced by rising solvent prices, government programs also coordinated these efforts.

Other moderately chlorinated wastes are being eliminated from at-sea activities due to West Germany's stringent program against highly chlorinated wastes. With federal prompting to recycle its highly chlorinated stream, Bayer AG announced that it would terminate all its ocean incineration activities by 1988. Unable to have their highly chlorinated wastes destroyed at sea, Bayer's managers chose to stop all at-sea burning. They will now combust their moderately chlorinated streams along with a sludge from one of their industrial wastewater plants. The termination of all ocean burning by Bayer shifts a 20,000–ton-per-year load from the sea to land-based treatment facilities.

Slightly Chlorinated Wastes: Destruction on Land

The slightly chlorinated wastes are generated in many of the same industries as the moderately chlorinated wastes, however, they are not reusable due to their low chlorine content. Instead they are destroyed in land-based hazardous waste incinerators with appropriate flue gas cleaning systems. West Germany feels that its special waste incinerators, such as the HIM incinerator, should receive these low-chlorine wastes. HIM has a chlorine capacity in its flue gas cleaning system of only 3,000 tons per year. This means that the capacity of the plant would be fully occupied for a year by just 5,000 tons of waste with a chlorine content of

60 percent. Moreover, the incineration of 1 ton of this highly chlorinated waste creates about 1 ton of hazardous waste from the neutralization of the flue gas.

In contrast, Table 4.5 shows that the total chlorine load of slightly chlorinated wastes is much smaller. Thus, despite the limited chlorine capacity of land-based hazardous waste incinerators, these installations can consume large amounts of the low-chlorine wastes. When the HIM hazardous waste incinerator was started up in 1982, about 10,000 tons of waste once burned at sea were rerouted to HIM by the German authorities. This set a trend. In the next few years, West Germany intends to employ similar land-based incinerators to completely replace ocean burning of their slightly chlorinated wastes.

CONCLUSIONS FOR AMERICA

In contrast to Germany's aggressive efforts to encourage waste reduction and land-based reuse of its chlorinated waste streams, America's regulatory framework appears incomplete and misleading. At present, it favors both waste reduction and ocean burning—and makes no effort at optimizing these programs in relation to each other. As a result, a number of American analysts are concerned that an increased use of ocean burning would seriously depress new developments in waste reduction and on-site treatment processes.

In a 1984 statement before the House Subcommittee on Fisheries and Wildlife, Jacques Cousteau warned against this course: "Do not allow fear and emotion to drive you like stampeding cattle to sea. There are alternatives available right now and we must go about using them right now.[18]" Considering the force of these concerns, EPA's reluctance to accept the Oslo Convention with West Germany's gusto is both destructive of and inconsistent with its current waste reduction campaign. Moreover, if the agency insists on its proposal to issue ten-year ocean incineration permits, it will guarantee an artificially inflated market by directly aborting innumerable small-scale competitors.

NOTES

1. H. U. Roll, "Welches Ausmass hat die Ölverschmutzung des Meeres?" in Umweltschutz (I): Wasserhaushalt, Binnengewasser, Höhe See und Küstengewasser–Zur Sache 3/71, 157–62.

2. Jernelov, S. et al.; "Biological Effects and Physical Properties in the Marine Environment of Aliphatic Chlorinated Byproducts from Vinyl Chloride Production," *Water Research* 6 (1117–91), 1985.

3. H. Wulf, "Verbrennung auf See bei Windstärke 9," *Umwelt*, March 1972, p. 48; H. W. Fabian, "Verbrennung chlorierter Kohlenwasserstoffe," *Umwelt*, January 1979, p. 12.

4. K.-H. Decker, "Verbrennung halogenhaltiger Abfalllösemittel auf Höher See," *Chemie-Technik*, 1980, p. 491.

5. Rat von Sachverständigen für Umweltfragen; "Umweltprobleme der Nordsee," 1980.

6. Gesetz zu den Obereinkommen vom 15. Februar 1972 und 29. Dezember 1972 zur Verhütung der Meeresverschmutzung durch das Einbringen von Abfallen durch Schiffe and Luftfahrzeuge vom 11. Februar 1977 (B L Bl. II, p. 165), and supporting documents, BT-Drucksachen 7/5268, 7/5610, 7/5640, H. von Lersner, Hosel v. Lersner: Recht der Abfallbeseitigung, Erg.-Lfg. V/79, Kz.1140, Rn 21.

7. E. Offhaus, "Prüfung der Beseitigungsmöglichkeiten an Land," in *Abfallbeseitigung auf See, Beihefte zu Müll und Abfall,* 1985, pp. 21–25.

8. Regulations for the Control of Incineration of Wastes and Other Matter at Sea, Addendum to annex 1 to the London Convention, 1978.

9. Technical Guidelines on the Control of Incineration of Wastes and Other Matter at Sea, London Convention, 1979.

10. Rules of Incineration at Sea, Annex 4 to the Oslo Convention, 1982; Code of Practice for the Incineration of Wastes at Sea, Oslo Convention, 1982.

11. U.S. Environmental Protection Agency, Science Advisory Board, *Report on the Incineration of Liquid Hazardous Wastes*, Washington, D.C., April 1, 1985; Pamela S. Zurer, "Incineration of Hazardous Wastes At Sea", *Chemical Engineering News,* December 9, 1985, p. 31.

12. "Untersuchungen zum Verbrennungswirkungsgrad und der Vernichtungsleistung bei der Verbrennung von ausgewählten Abfallstoffen auf einem Verbrennungsschiff," Bericht Nr. 79–10404149, TOV, Essen, 1980.

13. "Monitoring of Combustion Efficiency and Destruction Efficiency during the Certification Voyage of the Incineration Vessel "Vulcanus II," January 1983," Report R 83/53, TNO, Delft, 1983.

14. E. W. Kleppinger, and D. H. Bond, "Ocean incineration of hazardous waste: A revisit to the controversy" (consulting report) March 1985.

15. H. Sutter, "Tendenzen bei der Verbrennung von chlorierten Kohlenwasserstoffen auf See," *Müll und Abfall,* April 1984.

16. H. H. Fricke, "Zwangsanfall von Rückstanden und ihre Beseitigungsmöglichkeiten an Land," in *Abfallseitigung auf See, Beihefte zu Müll und Abfall* 17 (1980).

17. J. Schulze, and M. Weiser, "Vermeidungs- und Verwertungsmöglichkeiten von Rückstanden bei der Herstellung chlororganischer Produkte," Texte Umweltbundesamt, 1985.

18. Keith Scheider, "The Leper Ships," *Oceans,* May 1984, p. 69.

Part 2

Government Control of Risk

5

Government Ownership of Risk: Guaranteeing a Treatment Infrastructure _____

GARY DAVIS AND JOANNE LINNEROOTH,
WITH BRUCE PIASECKI

INTRODUCTION

Most Western, industrialized nations have hazardous waste regulatory systems for identifying hazardous waste streams, tracking their transport, and controlling and monitoring their management. Despite the ambition of these systems, government officials, industry, and the public express concerns that severe problems continue to exist in controlling the hazardous waste life cycle from its "cradle" to its "grave." In many countries this gap between the stated goals of the formal regulatory system and the actual results of the system can be attributed in part to the lack of a comprehensive industrial and economic infrastructure for handling the large volumes of hazardous wastes generated.

A whole spectrum of technical and policy options exist for the management of hazardous wastes. The type of system that develops in a country depends ultimately on the political and regulatory culture of the country, but also hinges on the perception of hazardous waste as a risk problem by the many actors involved. For instance, the issue may be framed by some groups as a problem of illegal dumping of wastes, in which case government ownership of waste management facilities with accompanying price subsidies may encourage legal management. For other groups the issue may emerge from environmental concerns over the large and increasing volumes of hazardous wastes being generated, in which case an aggressive policy of mandating high-priced technologies without subsidies or government involvement will encourage waste reduction—as well as illegal dumping. This tension between encouraging the legal disposal of wastes and promoting the reduction of wastes is one of many policy issues to consider in making investment and financial choices for waste-handling facilities. In making these choices, it is instructive to examine the accumulated experiences of those countries with established and functioning infrastructures for the management of hazardous wastes.

Two states in the Federal Republic of Germany (FRG), Hessen and Bavaria, along with the Scandinavian nations of Denmark, Sweden, and Finland, have developed unique systems of public financing and ownership of integrated hazardous waste management facilities that have condensed the hazardous waste life cycle into a comprehensive regulatory and management organization. These systems are in response to the intrinsic properties of hazardous waste as a policy issue discussed in Chapter 1. These political jurisdictions have large, integrated facilities for storage, treatment, and incineration of wastes that are equipped with state-of-the-art technologies. As a result, compared with other countries, relatively little of the hazardous waste generated within these areas is deposited directly in landfills; most is first pretreated or incinerated at high temperatures reducing both volume and toxicity.

In this chapter we will describe the history and operation of the hazardous waste management systems in Bavaria and Hessen in the FRG and in Denmark, Sweden, and Finland, contrasting these systems with more market-oriented approaches found, for example, in another state in the FRG, North Rhine-Westfalia, as well as in the United States. Our analysis of these contrasting management styles will form the basis for a discussion of the merits and drawbacks of public ownership, integrated facilities, monopoly markets, and subsidized pricing schemes, as well as how contrasting organizations have dealt with the inherent conflicts between promoting a market for capital-intensive facilities as well as motivating generators to reduce their wastes.

TECHNICAL OPTIONS FOR MANAGING HAZARDOUS WASTES

The predominant method for managing hazardous wastes in many industrialized countries remains land disposal. This includes landfills, where hazardous waste is placed in or on the ground; surface impoundments or ponds, where liquid waste is contained in natural or man-made depressions; land spreading or farming, where waste is tilled into the topsoil for biological degradation; and deep-well injection, where liquid waste is pumped underground into geologic formations thought to be sealed off from above. In the United States, for example, deep-well injection, surface impoundments, and landfills account for an estimated 80–90 percent of the off-site disposal of industrial hazardous wastes.[1] Many other industrial countries report similar figures.

Increased documentation of the risks from land disposal has led to growing interest in the development and promotion of alternative technologies and management strategies for hazardous waste. Based on the risks to human health and the environment, the following technological hierarchy is almost universally agreed upon as the best management strategy:[2]

1. Waste reduction or recycling: preempting the generation of hazardous wastes by process changes and reuse of valuable chemicals.

2. Physical, chemical, and biological treatments: these include physical processes, such as mechanical filtering, chemical processes by which the molecular structure of the waste is changed, and biological processes that rely on microorganisms to treat organic materials. These treatment techniques can render wastes innocuous or reduce their toxicity.

3. Thermal destruction: controlled thermal treatment, such as high-temperature incineration, destroys organic wastes and can be used to recover energy.

4. Solidification/stabilization of remaining residuals before disposal: this method consists of various techniques to "solidify" or encapsulate wastes to make them less likely to migrate when placed in a disposal facility. (More information is presented about technologies for hazardous waste treatment in Chapter 3.)

Bavaria, Hessen, and Denmark, in particular, stand out as having successfully implemented this preferred technical management hierarchy, reportedly placing few hazardous wastes directly into landfills. (The Swedish and Finnish systems have been operating for too short a time to draw any conclusions about their efficacy.) This stands in contrast to the United States and to other European nations as well, such as the Netherlands, where the majority of hazardous waste generated is sent to domestic and foreign landfills[3] and the United Kingdom, which disposes of 75 percent of its hazardous waste in landfills.[4] Also, in the rest of the FRG there is more reliance on direct land disposal for hazardous waste management than in Bavaria and Hessen.[5] Differences in definitions and reporting of hazardous wastes, however, make it difficult to accurately compare these figures among nations.

BAVARIA AND HESSEN: LEADERS IN THE FEDERAL REPUBLIC OF GERMANY

Background: The West German Political System for Waste Management

The Constitution of 1949 established the Federal Republic of Germany as a federation of what are currently eleven autonomous states, called *Länder*. Most regulatory power is distributed between the federal government and the states; however, in contrast to the United States, the Constitution of the FRG vests in the states the primary responsibility for implementing and enforcing most laws enacted by the federal parliament. The *Länder* therefore enjoy a more powerful position in relation to the federal government than their American counterparts.

Hazardous waste in the FRG is regulated under the Federal Waste Disposal Act of 1972, as amended in 1976, 1980, 1982, 1985, and 1986. As in other countries, the legislation was designed to cope with waste in general and not only hazardous waste. Certain types of waste have been regulated separately and excluded from the Waste Disposal Act, such as waste oil (see Chapter 6 for a discussion of recent changes to bring waste oil into the waste regulatory program), nuclear waste, waste-water, military wastes, and wastes from mining. Dumping at sea is regulated by the Dumping at Sea Act of 1977 discussed in Chapter 4.

The Waste Disposal Act and its accompanying administrative orders are formulated at the federal level and implementated by the states, in cooperation with the counties (*Kreise*) and municipalities (*Gemeinde*). Consistent with regulatory tradition in the FRG, this legislation lays out a framework which the *Länder* are obliged to follow, but which is general enough to allow them a great deal of discretion in choosing how the statutory goals will be met.

The federal legislation requires each generator of waste, household and industrial, to deliver his waste to the competent county or municipal authorities for treatment or disposal. At the same time, it assigns responsibility to the local authorities to handle the waste generated in their region by providing adequate facilities. With the consent of the respective state authorities, however, these local bodies can be relieved of their obligation to handle hazardous wastes, and most local bodies have followed this course.

In an administrative order of 1977, 86 specific waste types were listed as hazardous ("special") wastes. No procedure was specified for adding to or subtracting from this list, but the states have a right to supplement it in their own waste plans.

The Waste Disposal Act also lays out a "cradle-to-grave" control system, where a mandatory manifest procedure traces the path of each hazardous waste from its generation to its place of disposal. In addition, operators of certain facilities generating or handling hazardous wastes are required to appoint a waste disposal agent, protected from dismissal by law, who monitors the production, transport, and ultimate disposal of wastes.

A waste disposal plan must be drawn up by each state, and is binding for local and district authorities. The law requires that waste may only be treated, stored, or deposited in approved installations, and transported by certified transporters. The licensing of waste management facilities is generally done in regional land use permit proceedings, involving public comment and public hearings.

Although the regulatory system set up in the FRG is relatively comprehensive on paper, there are some weak points in the implementation of the system. To begin with, the effectiveness of the regulatory framework to "protect the well being of the general public" will depend crucially on whether all wastes that are in fact hazardous are included in the universe of regulated wastes. Some states, most notably Hessen, have greatly expanded the list of hazardous wastes.

Ways of legally evading the system also exist. A serious loophole has existed in the used oil exclusion, since used oil could legally contain a certain amount of hazardous substances and has sometimes been sold as heating oil for apartment buildings and houses. A second, equally important, loophole concerns the recycling of waste; generators may declare their waste as an economic good if they have a willing buyer, and in so doing are not required to enter it in the notification system. In addition, the legal possibility exists in the federal frame-work for generators to export their waste to other states or countries with less stringent requirements. Bavaria and Hessen, however, have imposed their own export restrictions in order to protect the market of waste for their own public treatment facilites.

These loopholes are considered to be serious, and attempts to close them by amending the Federal Waste Disposal Act have recently occurred. A recent amendment to the waste law is intended to discourage the export of hazardous waste from the FRG and to control imports.[6] As of November 1986 waste oil is subject to a testing requirement and certain contaminated waste oil is regulated as hazardous waste.

Other than the general framework laid out in the Waste Disposal Act and administrative orders, the federal government has little to do with regulating the actual practices of waste management. The *Länder*, assisted by the county and municipal authorities, have full responsibility for implementing the legislation. The practices of the states in this respect vary considerably.[7]

The discretion that the states enjoy in implementing the Waste Disposal Act is reflected most strongly by the Waste Disposal Plans, which show significant variations between states regarding the organization of waste management sys-tems and the type of disposal or treatment prescribed for different waste streams. A new federal administrative order is being developed by a working group consisting of state regulators and the Federal Environmental Agency (Umwelt-bundesamt [UBA]) to specify how each waste should be managed, with the goal of making waste management practices more consistent throughout the country (see Appendix 3 for excerpts).[8]

The main concern of this chapter is the strategic and organizational differences between those states that have publicly owned or publicly supported integrated hazardous waste management facilities and those that rely on a larger number of privately owned and decentralized facilities. These differences go beyond that of organization, since the outcomes of the systems themselves are very different. As mentioned above, Hessen and Bavaria place a relatively small proportion of their wastes directly in landfills compared to other West German states. In the next two sections, we will describe in some detail the Hessian and Bavarian waste management systems, including the technologies employed and their eco-nomic, legal, and institutional context.

Hazardous Waste Practices in Bavaria

Bavaria is considered by many waste experts to have one of the most successful systems for managing hazardous wastes in the world. This is due principally to its modern integrated treatment, storage, and disposal facilities, and to the gov-ernment's willingness to use its regulatory authority to ensure that generators utilize these facilities.

Bavaria is the largest of the West German states, with an area about the size

Table 5.1
Waste Management Methods in Bavaria

	GSB	%	ZVSMM*	%
Incineration	53,000 t	45%	18,000 t	14%
Chemical/Physical Treatment	97,000 t	29%	45,000 t	34%
Landfill	55,000 t	26%	70,000 t	53%

*The relatively large amount of wastes going to landfills is principally due to one generator who disposes of large quantities of wastes containing vanadium and chromium that cannot be treated economically.

of the state of Ohio in the United States and with nearly 11 million inhabitants. Yet it is not highly industrialized, producing less hazardous waste than, for example, Hessen and North Rhine-Westfalia. Around 417,000 tons of industrial waste were produced in 1983, 118,000 of which were considered hazardous waste as defined by the Bavarian waste list. The State has a relatively dispersed system of waste production, with approximately 6,000 hazardous waste generators and around 120,000 shipments of hazardous waste per year.[9]

Bavaria is of special interest because it recognized the hazardous waste problem very early and became a forerunner in improving hazardous waste practices, even significantly shaping the federal legislation. As early as 1966 the Bavarian district of Middle Franconia founded a municipal cooperative (Zweckverband Sondermüllplatze Mittelfranken [ZVSMM]) responsible for management of hazardous waste, and in 1970 a semi-public organization (Gesellschaft zur Beseitigung von Sondermüll in Bayern mbH [GSB]) was created to handle hazardous wastes for the rest of Bavaria. In both instances central facilities were built for treating and incinerating hazardous wastes and disposing of treatment residues.[10]

In 1983 228,000 metric tons and 117,000 metric tons of industrial waste, including hazardous waste, were managed by the GSB and the ZVSMM, respectively. The proportion of this waste that was incinerated, treated by chemical or physical methods, and deposited in secure landfills (mostly treatment residues) is shown in Table 5.1.[11] In addition, an unknown amount of hazardous waste is managed at approximately twenty industrial facilities where the generators have been given permission to handle their own waste.

The Bavarian Waste Law echoes the framework legislation of the Federal Waste Disposal Act of 1972. More detailed provisions governing special waste management are contained in the Bavarian Waste Plan, adopted in 1977. Waste generators must notify the Bavarian Environmental Protection Bureau of the Bavarian Ministry for Land Development and Environment of the types and quantities of waste they produce. The bureau then decides whether this waste should be managed as hazardous waste and in which categories it belongs.

The export of hazardous waste from Bavaria is prohibited without permission from the bureau, but exemptions from this prohibition can be secured for those wastes that cannot be burned, treated, or disposed of safely in Bavaria or when export may be justified by the ability of a large company to treat waste at its own facility in another state. Generators of hazardous waste must obtain permission from the bureau to manage their wastes on-site. Very large chemical companies such as Hoechst, for example, burn their own waste. This practice, however, is generally discouraged, and stricter environmental standards are often applied to on-site facilites than to the GSB and ZVSMM plants. In 1983 approximately 9,200 tons of industrial waste were exported to other states in the FRG for management.[12]

ZVSMM: Local Government Initiative

Concern about groundwater contamination from the dumping of hazardous waste in municipal waste dumpsites inspired politicians in the District of Middle Franconia to form the Zweckverband Sondermüllplatze Mittelfraken (ZVSMM) in 1966, at which time there was no effective legislation in the FRG for dealing with the disposal of hazardous waste (for more information on ZVSMM and the Schwabach facility, see Chapter 3). The ZVSMM is of special interest in that it is a fully public enterprise, in contrast to the joint government/industry-owned facilities in the rest of Bavaria and in Hessen. It is owned and operated by the municipal governments within the district.[13]

ZVSMM owns and operates the Schwabach facility, which consists of a rotary kiln incinerator for organic waste, a physical/chemical treatment plant, a wastewater purification plant, and a clay-lined hazardous waste landfill (which includes leachate collection and treatment). ZVSMM has also recently opened another landfill site at Raindorf, approximately twenty kilometers from Schwabach. These facilites serve, in addition to Middle Franconia, other parts of Bavaria and the neighboring state of Baden-Württemberg (about 10 percent of the area of the FRG), 3.7 million inhabitants, and approximately 4,000 industrial companies. The ZVSMM does not operate collection vehicles, but relies on private transportation companies.[14]

The Schwabach facilities are financed by ZVSMM and subsidized by the Bavarian state government. As of 1985 a total of DM 44.5 million (about $23 million) had been invested in the facilities. The members of ZVSMM—five large towns, seven county districts, and seven small towns—raised DM 400,000 (about $200,000) of the original capital investment of DM 5 million (about $2.5 million). The Bavarian government added approximately DM 2 million (about $1 million) in the form of a grant, and the remaining funds came from loans (including a DM 1 million low-interest loan from the Marshall Fund). Subsequent construction has received a 30–50 percent subsidy from the Bavarian government that will not have to be repaid.[15]

The state government, therefore, subsidizes the ZVSMM facilities directly by contributing to the capital investment. The fees charged to generators for management of waste at Schwabach cover the remaining costs, including the non-subsidized capital costs, operating expenses, and interest payments. ZVSMM operates on a non-profit basis, and prices are set at the end of the year to reflect the anticipated costs for the next year. Prices for 1985 ranged as follows:[16]

Landfill	DM 65–195 per ton	($32–97)
Treatment	DM 70–550 per ton	($35–225)
Incineration	DM 80–580 per ton	($40–290)

At the direction of the district of Middle Franconia, ZVSMM is currently installing a new flue gas scrubber for the incinerator that will result in a price increase of approximately DM 100 per ton (about $50) for incineration.[17]

GSB: Government-Industry Partnership

The GSB manages hazardous wastes in the remainder of Bavaria. The state waste plan specifies that all hazardous wastes in Bavaria other than those handled by ZVSMM must be delivered to GSB unless permission is granted for on-site disposal or export. GSB has four treatment or disposal facilities, and seven transfer stations in Bavaria consolidate waste for shipment to these facilities. These transfer stations also dewater sludges and oil/water emulsions and perform some acid-base neutralization to reduce volumes before shipment.[18]

The Ebenhausen plant, the main facility for hazardous waste treatment, located about 50 kilometers from Munich, includes two rotary kiln incinerators, a physical/chemical treatment plant, and a wastewater treatment plant. The Schweinfurt disposal plant consists of an incinerator for certain industrial wastes, such as paper contaminated with oil, which can be incinerated with domestic refuse. GSB also operates a solvent-recycling facility near Munich and a large landfill at Gallenbach. (The Ebenhausen facility is described in greater detail in Appendix 1.) For the purposes of this chapter, it is important to note that most hazardous waste containing organics is incinerated and most toxic inorganic waste is treated to reduce toxicity and mobility. The landfill does not generally accept liquid waste or any waste that has not been pretreated.[19]

The original capital stock for the GSB facilities of DM 1 million (about $500,000) was raised by 76 hazardous waste generators (30 percent), the Bavarian government (40 percent), and member communities (30 percent). The original DM 1 million of GSB member stocks had risen to a total of DM 21 million (about $10.5 million) as of 1980. Bavaria now has an interest of 78 percent, industry 14 percent, and the municipalities 8 percent. The remaining outlays have been financed by a combination of direct government subsidies, indirect subsidies of low-interest government loans, and user fees.[20]

Prices per ton for hazardous waste management for 1985 ranged as follows:[21]

Landfill	DM 64–195	($32–97)
Treatment	DM 70–490	($35–245)
Incineration	DM 115–620	($57–310)

Hazardous Waste Practices in Hessen

Hessen also has a comprehensive hazardous waste management system. Emphasis is put on waste reduction; most hazardous waste generated is incinerated or treated; and little hazardous waste is deposited in landfills without prior treatment.

According to the Hessian Ministry of the Environment, approximately 300,000 tons per year of hazardous waste are currently generated in Hessen and sent off-site for management.[22] This is out of a total of 700,000 tons of solid waste from industry.

Hessen has devised a degree-of-hazard system for the management of all waste on its waste list and has greatly increased the number of waste types requiring treatment as hazardous waste. Waste is divided into three categories, each with a specified management method:[23]

I. Industrial waste that is similar to municipal garbage and can generally be treated as such

II. Industrial waste that is hazardous and cannot be disposed with household waste, thereby requiring special handling, such as treatment, incineration, or deposition in secure landfills

III. Industrial wastes that are especially hazardous and require treatment under "special technical conditions," such as salt mine disposal or high temperature incineration.

Categories II and III can be regarded as hazardous waste, where category III waste is special priority or especially hazardous. From the 563 waste types listed in the federal catalog, 262 are identified as category II and 37 as category III. This is compared with 86 hazardous waste types in the federal list.[24] For the approximately 300,000 tons of hazardous waste in categories II and III, Table 5.2. shows how they were managed in 1983. Much of the waste deposited in secure landfills was residues from incineration and physical/chemical treatment, and a large portion of the waste was sent to other states for disposal, since Hessen lacks landfill capacity.

The organization of hazardous waste management in Hessen is legally based on the Federal Waste Disposal Act and administrative orders, the Hessian Waste Law, adopted in 1978, and the Hessian Waste Management Plan. This plan assigns full responsibility for special or hazardous wastes to the Hessische Industriemüll GmbH (HIM), which was established in 1974. Originally the HIM was a private waste management company financed by hazardous waste generators, but when it encountered financial difficulties, the Hessian government

Table 5.2
Waste Management Methods in Hessen

Incineration	43,000 metric tons
Chemical/Physical Treatment	100,000 metric tons
Landfill	136,000 metric tons
Underground Deposit	13,000 metric tons

Source: B. Furmaier, "Organisation und Stand der Sondermüllbeseitigung in Bayern,"
    unpublished paper, October 1983.

rescued it and assumed joint responsibility for the facilities. The HIM is currently jointly owned by the Hessian government (26 percent) and a consortium of 25 Hessian waste-producing industries (74 percent). The Hessian government has three votes on the board of HIM, compared to industry's eight.[25] Recent political changes in Hessen, with the Green party becoming part of a governing coalition, may very well lead to the Hessian state government assuming a controlling interest in HIM.[26]

Government authorities in Hessen thus exert a direct control over hazardous waste management through their part ownership of HIM, and indirect control through regulation. The responsible agency is the Hessian Ministry of the Environment, which categorizes waste, specifies how it will be managed and promulgates standards for emissions from industrial facilities. The regional authorities are primarily responsible for monitoring and regulating the transportation of waste and the operation of waste facilities.

HIM operates four hazardous waste facilities in Hessen. The most recently constructed is the Biebesheim incineration facility, completed in 1981. Two physical/chemical treatment plants are in operation in Frankfurt and Kassel, and a small landfill exists for the district of Marburg.[27] A large landfill was planned for Mainflingen, but encountered extensive public and political opposition and will probably not be constructed as planned.[28] In addition to these HIM facilities, Kali and Salz AG operates an underground salt mine deposit near Herfa-Neurode that accepts hazardous waste for long-term storage (see Chapter 3 for more details).

Transfer stations for hazardous waste do not exist in Hessen, but several are planned, and the HIM has been testing a waste pickup service for small generators to enourage the safe management of small quantities of hazardous waste. Industrial customers that generate less than 500 kilograms per year can deliver their waste to a special truck and pay only DM 1 per kilogram. Citizens bringing household toxic wastes may do so free. The system is financed by a tax on waste production, its level depending on the quantity and type of waste produced.[29]

The Biebesheim incineration facility consists of two rotary kilns, afterburners, heat recovery, and a novel scrubbing system for the exhaust gases. The physical/ chemical treatment plants at Frankfurt and Kassel use standard technologies for cyanide destruction, neutralization, and precipitation. Only solids or dewatered sludges, usually treatment residues, are permitted to be landfilled in Hessen. Lacking landfill capacity, HIM exports most of the treatment residues and other waste for landfill to neighboring states. (Baden-Württemberg, Bavaria, Lower Saxony, and North Rhine-Westfalia). Only the most toxic and persistent waste that cannot be easily treated or safely disposed is sent to the salt mines at Herfa-Neurode.[30]

Hazardous waste may be treated, incinerated, or disposed of on-site by the generator only with special permission from HIM; statistics on the amounts handled on-site are not available. Existing on-site disposal facilities were permitted to continue to operate, but no new facilities with the exception of wastewater treatment have been permitted. Firms handling their own waste are generally not permitted to accept waste from other generators. Except with special permission, the export of hazardous waste from Hessen is forbidden, and the export of treatment residues and other waste to landfills outside Hessen is performed only by HIM.[31] The restrictions on competition and export protect the economic viability of the HIM facilities.

Regional authorities in Hessen, which have licensing authority for all industrial facilites, are encouraging waste reduction by requiring documentation of waste reduction and recycling measures by those seeking permission from the planning authorities to construct new industrial facilities and to expand existing ones. The authorities can deny this permission unless the facility owner includes up-to-date measures for pollution reduction and recycling processes. Due to public scrutiny of new industrial facilities, some facility owners have had to produce extensive documentation concerning their efforts to reduce pollution and have been forced to make modifications in their processes. (See Chapter 2 for more details).[32]

The original industrial owners of HIM financed the first facilities, but cost difficulties forced the government of Hessen to become increasingly involved, first by supplying low-interest loans and later by directly subsidizing capital expenditures. The Hessian government paid most of the capital costs for the Biebesheim facility (DM 100 million, about $50 million) and recently contributed an additional subsidy of DM 9 million (about $4.5 million). The prices per ton for hazardous waste management by HIM in 1985 were as follows:[33]

Landfill	DM 120–360	($60–180)
Physical/chemical treatment	DM 60–660	($30–330)
Incineration	DM 400–3000	($200–1500)
Disposal in salt mines	DM 180	($90)

Because of the subsidies, these prices do not reflect the full costs of disposal.

The Hessian authorities are becoming concerned about underutilization of the facilities, especially the Biebesheim facility. In order to fully utilize the 60,000-ton-per-year capacity of Biebesheim, about 17,000 tons of waste per year from other Länder, such as Baden-Württemberg, and from other countries, were imported in 1983. In addition, 20,000 tons per year are imported to the Herfa-Neurode salt mines.[34]

THE DANISH SYSTEM; COLLECTIVE GOVERNMENT RESPONSIBILITY

Hazardous Waste Management in Practice

Denmark is a small country, about half the size of the state of Maine and twice the size of the state of Massachusetts, with a population of 5.1 million. Industry within Denmark is privately owned and highly developed, characterized by small and medium-sized firms. A chemical plant with 500 employees would be considered very large. Hazardous waste is called "chemical waste" in Denmark. All chemical waste is required to be managed in the most environmentally sound manner, regardless of whether it contains certain listed chemicals or has certain concentrations of hazardous constituents.[35]

The types of hazardous waste that are generated in Denmark are similar to those in other industralized countries. The total hazardous waste generation has been estimated at 100,000–150,000 metric tons per year by one authority, and approximately 100,000 tons per year by another.[36] Generators of hazardous waste are generally required to deliver their waste to collection and transfer stations owned by the municipalities for transport to the central treatment facility, Kommunekemi. If they obtain permission, there are three other options:

1. Waste can be recycled, treated, or disposed of on-site.
2. Waste can be sent to an approved recycling plant or sold to another industry through the waste exchange if a permit has been obtained specifically for the waste in question.
3. Waste can be exported to another country.[37]

The central treatment facility, the Kommunekemi plant, is located in approximately the geographical center of the country, on the island of Funen in the city of Nyborg. A network of 21 transfer stations all over the country feed industrial hazardous waste to the plant, mostly by rail car.[38]

The plant, which commenced operation in 1975, consists of three incinerators for organic wastes (including two large rotary kilns), an inorganic treatment plant, and a waste oil reclamation plant.[39] A landfill is located nearby, which receives treatment residues from Kommunekemi (slag from the incinerators and filter cakes from the inorganic treatment plant) and from industries treating wastes on-site. A very small amount of untreated wastes is deposited there (approximately 1,000 tons per year).[40] (For more information on the Kommunkekmi facility, see Chapter 3.) The prices per ton for waste management at Kommu-nekemi in 1985 were as follows:[41]

Incineration	Kr 40–4,500	($14–450)
Physical/chemical treatment	Kr 1,050–3,200	($105–320)
Land disposal	Kr 375–575	($33.50–57.50)

Certain wastes are not normally treated at Kommunekemi. These are PCBs, highly concentrated cyanide metal-hardening salts, organics containing high con-centrations of heavy metals, and old mercury batteries. This waste, totalling about 1,000 tons per year, has been sent to the Herfa Neurode salt mines in West Germany for disposal. There are also two firms in Denmark that recycle about 1200 tones per year of spent solvents. The still bottoms are sent to Kom-munekemi for incineration.[42]

The remainder of the waste that is generated in the country is either managed on-site or exported. There is very little overall information concerning the manner in which this is managed, since the municipalities grant permission for on-site management and there is no reporting to the national government required. The National Agency for Environmental Protection has recently been reviewing the exemptions granted by the municipalities to determine if the waste is being managed in facilities that are protective of the environment.

The Political Structure of Hazardous Waste Management

Although the management of hazardous waste is centralized, the regulation of this waste in Denmark is decentralized. Denmark has a parliamentary system of government, with national environmental affairs handled by the Ministry of the Environment. Within the Ministry of the Environment is the National Agency of Environmental Protection, which is generally responsible for pollution control. Denmark is divided into 277 municipalities and 14 counties, each with elected councils. It is these local governments that are ultimately responsible for en-forcing the hazardous waste laws and regulations and for approving new industrial operations that will generate hazardous waste.[43]

The Danish system of hazardous waste management was established in the early 1970s by the initiative of local governments throughout the country who perceived the need for effective hazardous waste management to prevent the disposal of hazardous waste in municipal garbage landfills. This may have been motivated by the fact that 98 percent of drinking water in Denmark comes from groundwater.[44] The hazardous waste management company, Kommunekemi, was set up as a public non-profit corporation by the National Association of Danish Municipalities in 1971, and the plant at Nyborg was constructed with a no-interest, deferred-payment loan from the Association of Municipalities. The municipalities played a key role in the development of the 1972 chemical waste legislation that gave them the principal authority for hazardous waste collection and management. The Act on Disposal of Oil Waste and Chemical Waste, passed in 1972, gave the Minister of Environment administrative authority to issue regulations pertaining to this waste. Aside from issuing these general regulations for hazardous waste, the Ministry of the Environment and the National Agency of Environmental Protection decide appeals from decisions made on the local level.

The municipal councils have the majority of the duties under the waste law. The county councils have a supervisory role over the municipal councils. The municipal councils are required to

1. establish mandatory local collection and receiving arrangements for waste oil and chemical waste;
2. receive notifications from all generators of waste oil and chemical waste within their jurisdictions;
3. approve exceptions from the mandatory delivery of wastes to the municipal collection or receiving arrangement;
4. enforce the regulations for waste management; and
5. issue instructions for cleaning up pollution of the environment.[45]

The universe of regulated "chemical waste" is defined by a list of 51 types of chemical waste that are known to be generated in Danish industry and that have general characteristics of toxicity, corrosivity, or flammability. A catchall phrase also brings in other types of waste that are similar to those listed, since it is impossible to anticipate new types of waste that will be generated. The general philosphy of the NAEP is that all waste oil and chemical waste that can be easily collected must be collected. There are no detailed criteria or lists of compounds and threshold concentrations for defining those waste types that are subject to the regulatory system.[46]

All generators of chemical waste must notify the municipalities of the types and quantities of chemical waste they are generating. To assist the municipalities in their task of ensuring that all generators notify, the NAEP has prepared a chemical waste handbook listing the types of waste that are generated by most industrial processses. The municipal council decides whether or not wastes that are not on the list of 51 types of chemical waste are subject to regulations.[47]

Generators of chemical waste must deliver this waste to a collection or transfer station selected by the municipal council, or the municipal council can institute a pickup service for the waste. The collection system was specified in the legislation and was set up on a regional basis over a period of three to four years through the cooperation of the municipalities. The costs of constructing the smaller collection stations and the central transfer stations were borne by the municipalities and shared proportionately according to population or waste generation. Each transfer station is now in the process of establishing a collection service, since the experience of those few stations that have provided the service has shown that the quantities of waste received are greater if the service is provided. The cost of transport to the transfer stations will be borne by the public through taxes, since the existing tax system is much easier to administer than a new waste fee system.[48]

If a generator can prove that he is capable of treating or disposing of his waste on-site in an environmentally acceptable manner, the municipal council must grant that generator an exemption from the obligation to deliver the waste to the municipal collection system. An exemption can also be obtained if the generator intends to have the waste recycled or treated by a private firm, or to sell it for use by another industry, or to export the waste to another country. The generator must make a showing that the facility to which it proposes to send the waste is authorized to receive that type of waste and will manage it in an environmentally sound manner. However, for waste that is exported from the country, there is little scrutiny about the manner in which it will be managed. The decisions regarding exemptions are appealable to the NAEP by either the affected generator or affected citizens. The NAEP's decision can be appealed to an Environmental Appeals Board.[49]

Probably the biggest potential for loopholes arises from the fact that different municipalities have different levels of enforcement of the notification and delivery requirements, and that the NAEP's role in monitoring the performance of the municipalities is more advisory than anything else. The fact that export permission is relatively easy to obtain has also made the capacity of the Kommunekemi facility underused.[50]

SWEDEN: STEPS TOWARD GOVERNMENT OWNERSHIP

Sweden is also a highly developed industrialized nation. Swedish industry and business generates an estimated 400,000 to 500,000 tons per year of hazardous waste and waste oil. Estimates by the Swedish Water and Air Pollution Research Institute (IVL) are that approximately half of this waste is managed on-site.[51] For some of the waste that needs to be managed off-site the Swedish government has established a public company for hazardous waste management, called SAKAB (Svensk Avfallskonvertering AB). SAKAB currently manages about 60,000 tons of waste oil and hazardous waste per year. Approximately 50,000 tons per year are managed by municipally owned treatment or disposal facilities. The remaining waste generated is predominantly waste oils and spent solvents and is recycled by privately owned facilities. Overall information on hazardous waste generation and management is difficult to obtain because the notification and manifest system is only now being implemented.[52]

Prior to the establishment of SAKAB, hazardous waste in Sweden was managed by private firms, and there are still at least twenty private firms that are permitted to recycle waste oil and spent solvents. Generators of hazardous waste that is sent off-site are required to deliver their waste to one of SAKAB's six transfer stations or to the central treatment facility unless an exemption is granted.[53]

The national government owns 96 percent of SAKAB, 2 percent is owned by Swedish industries who generate the hazardous waste, and 2 percent is owned by the Swedish municipalities. Similarly to the Danish model, SAKAB has constructed a centralized hazardous waste incineration and treatment facility and a network of transfer stations. The SAKAB Norrtorp Plant, which began operating in 1984, consists of one rotary kiln incineration system, an oil recovery plant, a physical/chemical treatment unit, a mercury recovery unit and an above-ground landfill area for treatment residues. The only waste, other than treatment residues from the plant, that is permitted to be disposed of in the landfill area is heavy metal sludges from treatment processes located at the generators' plants.[54]

The original capital stock for SAKAB was about Kr 28 million (about $4.2 million). Capital investments in the facility to date have totalled over Kr 250 million (about $37.5 million), which included a subsidy of Kr 43 million (about $6.5 million) from the national government to pay for an interim storage facility and a siting study, a loan of Kr 82 million (about $12.3 million) from the government with no interest and no payments for five years, and a loan of Kr 110 million (about $16.5 million) from Swedish banks with the government as a guarantor. The loans will presumably be paid back through the operating revenues of the facility.[55]

FINLAND: SHARED OWNERSHIP IN SCANDINAVIA

The Finnish hazardous waste management system is the newest national system to be created and is based upon the experience of the Danish and West Germans. Hazardous waste is called "problem waste" in the Finnish Waste Management Act, and includes waste oil. The estimates of total generation of hazardous waste and waste oil in Finland are 100,000 tons per year. On-site treatment accounts for approximately 30,000 tons per year of this total.[56]

Problem waste in Finland is managed by the Finnish Problem Waste Management Company (Oy Suomen Ongelmajate). A central treatment facility, which became operational is 1985, has been established near Riihimaki, a city of about 24,000 inhabitants in southern Finland. The preliminary design of the facility was performed by the Danish company Chemcontrol, of which Kommunekemi, the Danish public hazardous waste management company, is part owner. The facility includes one rotary kiln incineration system with a capacity of about 35,000 tons per year, a physical/chemical treatment unit with a capacity of about 3,000 tons per year, a wastewater treatment plant for emulsions and water contaminated with organics having a capacity of about 40,000 tons per year, and a landfill for treatment residues that has a capacity of about 15,000 tons per year. The incineration unit generates electricity that is used in the plant and sold to the national grid and steam that is delivered to the district heating system for Riihimaki.[57]

In the mid-1970s the national government of Finland began investigating hazardous waste generation and management and concluded in 1978 that the Finnish governnment should establish a quasi-public hazardous waste management system. Under the Waste Management Act, passed in 1978 and amended in 1982, municipalities have responsibility for the collection of waste oil, but hazardous waste transport to the proper facilities is the responsibility of the Problem Waste Management Company. A permit must be obtained to treat or dispose of hazardous waste on-site, and export of hazardous waste may be prohibited by the Minister of the Interior.

The Probelm Waste Management Company was founded in 1979 as a non-profit corporation by the national government, a consortium of municipalities, and Finnish industries that generate the waste. Each of the three parties owns one-third of the stock in the company. The company is governed by an eighteen-member Administrative Council with six representatives from each sector and a national government representative as the permanent chairman. The original capital stock was about $1 million and the capital costs of the facility were approximately FIM 265 (about $66 million). The capital for construction of the facility was mainly provided by loans guaranteed by the national government, although the government contributed about FIM 45 (about $11 million) as a subsidy. Except for the subsidies, the capital costs are being recovered by charges for waste management.[58]

The Finnish government cited the following advantages as its reason for establishing a centralized, quasi-public facility, with transportation performed by the same organization:

1. The economy of scale of having a single large facility to treat all of the nation's hazardous waste.
2. Greater ability to supervise the management of hazardous waste since it would all be managed in a single facility under partial government ownership.
3. Supervision of waste transport would be easier, minimizing the risks of transportation.
4. The incineration facility could produce enough energy for recovery and use for other purposes.

POLICY STRATEGIES FOR A MANAGEMENT INFRASTRUCTURE

Policy Objectives

The management of hazardous waste in an environmentally acceptable way with first priority on the reduction or recycling of waste and minimization of direct land disposal is a major environmental goal in the United States. This objective is generally agreed upon by governments, environmental groups, and industry. Yet inherent in even this seemingly straightforward technical hierarchy are some fundamental strategic conflicts.

First, policies that encourage generators to reduce their waste by making disposal difficult and expensive will simultaneously and inevitably lead generators to find loopholes in the regulations and possibly even to dispose of their waste illegally. Especially where voluntary compliance is so important, many argue for the cooption of industry through economic incentives or lower prices. Others argue that industry should pay the full social costs of its polluting behavior and that the taxpayer should not subsidize the management of hazardous wastes.

The tension between promoting the eventual reduction of hazardous waste and ensuring the immediate goal of safe treatment and disposal is complicated by a second contradiction arising from the need to create a waste market to sustain the capital-intensive treatment and disposal investments. The high costs of these investments, if passed on to the generator, will ultimately force the reduction of wastes and consequently undermine the financial viability of the facilities. Private investors are understandably reluctant to enter a declining market. In sum, the long-run reduction of hazardous waste, which is the most environmentally sound waste management strategy, is in some ways incompatible with shorter-run needs to create and maintain an expensive treatment and disposal infrastructure to handle the large amounts of waste presently generated.

This tension leads to the following four sometimes conflicting management goals:

1. The management of currently generated hazardous wastes in the most environmentally sound ways by promulgating a comprehensive control system and establishing a technical and economic management infrastructure
2. The reduction of hazardous waste generation
3. The promotion of full compliance with the control system by eliminating illegal practices
4. The allocation of the full costs of hazardous waste management to the generators

Most national regulatory systems address primarily the first and third of these goals by setting up a control system to track waste from "cradle" to "grave" to ensure proper handling by licensed facilites. The success of this control system rests on the existence of a network of management facilities with sufficient capacity, a functioning transportation system, and knowledgeable personnel. Policies for establishing this infrastructure, which are the focus of this chapter, range from full government support in financing and operation of facilities to more laissez faire strategies that depend on the initiative of private entrepreneurs.

The second goal of reducing the generation of hazardous waste can be promoted through various policy measures including high prices for mandatory treatment or disposal methods and other economic incentives, such as a tax on waste (waste-end tax) or subsidies for recycling waste, the financing and operation of a waste exchange, government-supported innovation to find alternative production processes, or an outright ban on the generation of certain very hazardous wastes (see Chapter 2 for a detailed discussion of waste reduction policies).

Voluntary compliance with the system by the elimination of illegal practices, the third goal, might be promoted through lowered waste management prices for generators, made possible by public subsidies, or by making the generators clearly liable for any environmental or health damage resulting from the management of their waste. Alternatively, the regulatory authorities can force compliance with more frequent inspections and other enforcement measures combined with formidable sanctions.

The fourth objective of allocating the full costs of hazardous waste management to the generators of waste, or adherence to the "polluter pays" principle, would mean eliminating any public subsidies that are passed on in lowered prices to the generators. It would also require measures to guarantee that industry pays the full social cost of unexpected pollution, such as an industry-financed post-closure fund for hazardous waste facilities and other funds to cover pollution damage, as well as compulsory insurance or other proof of financial responsibility to pay for damages.

Since the focus of this chapter is on policy choices for establishing an economic and institutional infrastructure for the treatment and disposal of hazardous waste in an environmentally acceptable way, or the implementation of the first goal, policies to achieve the other three goals will not be discussed in detail. Nonetheless, the inevitable trade-offs between the first goal and each of the others will be highlighted.

Organization Options for a Treatment Infrastructure

The governments discussed in the first part of this chapter have built technologically advanced, integrated facilities for the storage, treatment, and disposal of hazardous waste with a significant investment of public funds, placed a high priority on relatively expensive treatment and incineration technologies with little direct land disposal, have required generators to use the publicly owned management facilities in monopolisitic fashion, and have shared the cost of hazardous waste management between industry and the taxpayer. This "public monopoly" strategy stands in direct contrast, for example, to the United States, which has relied on the private market to provide a network of management facilities with an emphasis on land disposal and generally with no public subsidies. In between these two contrasting systems are many diverse economic and institutional possibilities for creating a management infrastructure where the four key variables are the following:

1. The extent of public versus private ownership and control
2. A monopolized versus competitive market
3. Integrated, comprehensive facilities versus segregated, specialized facilities
4. Cost allocation between industry and the public

These four fundamental policy variables are illustrated by the "policy tree" shown in Figure 5.1. The most basic question concerns the extent to which the government assumes responsibility for hazardous waste management through such means as public ownership or sponsorship. The second related question concerns the type of economic organization of the facilities where the possibility exists for creating and sustaining a public or private monopoly by restricting entry of other firms and assuring a market for the facilities through such means as export controls. A third branch of the "policy tree" represents the choice between integrated and specialized facilities, with the caveat that the distinction is not clearly demarcated nor is the choice for or against integrated facilities wholly separate from the choice for or against a monopoly investment venture. A fourth choice or branch on the "policy tree," again not wholly separate from the other choices, concerns who ultimately pays the costs of hazardous waste disposal—industry or the taxpayer. The government can subsidize treatment and disposal facilities through such means as investment credits, lowered interest rates on capital investments, tax breaks, or more straightforward grants to cover capital costs and operating expenses.

As shown in Figure 5.1, various combinations of these policy variables can lead to at least fourteen different forms of economic and physical organization for a treatment and disposal infrastructure. The two extremes in terms of government intervention are represented, on the one hand, by Denmark and Bavaria with their publicly owned monopoly facilities, which are integrated and operate with large public subsidies, and, on the other hand, by states in the United States with privately owned, competitive, and specialized facilities that generally operate with no public money. As illustrated, the Danish and Bavarian systems are one of four possible variations of what we call the public monopoly model. The U.S. model is one of four variations of what we call the private competitive model. In between, there are different combinations of private and regulated monopolies, which we call the public utility model.

Public versus Private Ownership: The First Branch of the Policy Tree

The most striking feature of the examples discussed in the first part of this chapter, especially in comparison with the United States, is the extent of public ownership of the waste disposal facilities. By financing capital investments and taking responsibility for operations, the governments discussed have, in effect, assumed responsibility for the risks of hazardous wastes. The liability and responsibility of the generator for the long-term disposal of hazardous waste ends once it is ceded to the public authorities. This contrasts with the principle of generator responsibility in other countries, including the United States.

From the experiences of the countries discussed above, we find that a strong motivating force for public involvement was either to finance the large capital investments needed for centralized incineration and treatment facilities or to

Figure 5.1
A "Policy Tree" for Establishing a Treatment Infrastructure

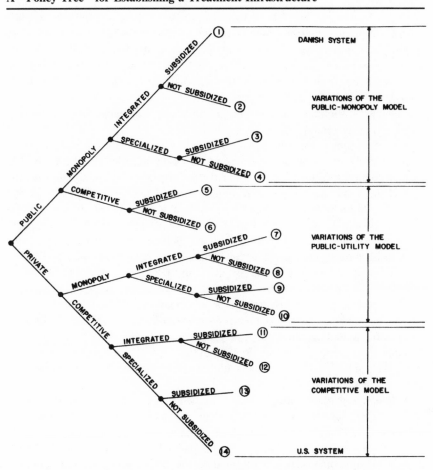

cover the losses these facilities incurred once in operation. The gradual slide toward public ownership, for instance, of the GSB and HIM facilities in the Federal Republic of Germany, a country with a strong free-market ideology, was viewed positively by these states as the best means to promote environmental objectives.[59] The first argument for public management was tied to public subsidies, since such subsidies encourage proper waste management by allowing services to be provided to industry at "affordable" prices. As is apparent from Figure 5.1, several institutional arrangements can be coupled with public subsidies to keep prices lower, but governments providing subsidies naturally tend to assume more control over the activities subsidized to ensure that public money is used to satisfy public objectives.

The rationale for greater public control of the waste management facilities in

Bavaria and Hessen and in other countries, however, went beyond that of trans-
ferring some of the disposal costs to the taxpayer through public subsidies.
Among many waste officials in these countries there is a deeply entrenched
mistrust of private enterprise in the hazardous waste business, rooted in a history
of bad experiences.[60] Misgivings about the ability of the private market to handle
hazardous waste in the most environmentally correct way and to maintain the
long-term security of hazardous waste land disposal facilities were thus strong
motivating factors for public control. Public ownership in the countries studied
pushed the development of more environmentally acceptable, albeit more expensive, management techniques.

Another argument for public ownership and control, which was part and parcel
of the underlying philosophy that hazardous waste management is for environmental control and not a profit-oriented enterprise, was the obligation on the
part of the public authorities to accept all waste delivered to the facilities and
not select those with the highest economic value. Authorities in all of these
countries feared that the private market would not have the flexibility and breadth
to serve the whole range of waste generators, including those producing "dirty"
waste with little or no fuel value for incineration. A public enterprise with an
"environment before profit" philosophy could take this and spread the losses
incurred.

The reservations about private enterprise serving environmental needs can, in
theory, be countered by appropriate regulatory measures and by requiring post-
closure funds to cover the costs of long-term maintenance of land disposal
facilities. The "dirty" waste could also, in theory, be accommodated by private
enterprise, although at substantially higher prices that might create greater incentives for illegal disposal.

Although a regulatory program may, in theory, serve the same environmental
needs, experience in many countries has shown the difficulties in planning,
permitting, and monitoring private disposal facilities, which are complicated by
a lack of funds for hiring the requisite personnel as well as the general non-
availability of people sufficiently trained for the jobs involved. Central, publicly
owned facilities concentrate available expertise and make it easier to ensure that
management decisions are made with environmental goals as a priority. On the
down side, the absence of competition and profit motive for these facilities may
also lead to the well-known tendency for publicly owned industries to operate
inefficiently.

Essentially, the choice between private and public facilities can be viewed as
a trade-off between the inefficiencies of public enterprise and the substantial
difficulties in monitoring private enterprise to assure that environmental objectives are met. Public ownership and control of the facilities for treating and
disposing of hazardous waste thus has an apparent advantage over private ownership in that the control of environmental measures is more direct. But this
advantage hinges critically on the experience and consciousness of the civil
servants in charge, especially since they may be subject to fewer external checks.

The problems of industry acceptance of public management can be lessened if generators become partners in the waste management venture as in Bavaria, Hessen, Sweden, and Finland. Industry cooperation can serve as a valuable policing influence within the community of generators. This cooperation between industry and government in Bavaria was an essential ingredient that allowed a consensus to be reached quickly concerning the establishment of a hazardous waste management system, and the GSB was founded after only a little more than a year of negotiations.[61]

A closely related alternative to both the public enterprise and the hybrid public-private enterprise is the public utility model shown in Figure 5.1. Here the management facilities are privately owned, but closely regulated, and their market is guaranteed.

Monopoly versus Competition: The Second Branch of the Policy Tree

The European governments discussed in the first part of the chapter have created systems for disposing of hazardous waste that, by emphasizing treatment and incineration over direct land disposal, have reduced the environmental risks of hazardous wastes far below that of neighboring states and countries. Although the costs are spread between the public and industry, hazardous waste management is still more expensive in these jurisdictions than in those where land disposal predominates. Even considering transportation costs, it would often pay a generator to ship waste across the borders for land disposal rather than to in-state treatment and incineration facilities. To preserve the market for their treatment facilities, the authorities in these jurisdictions have forbidden the export of hazardous wastes without special permission and required all generators shipping their waste off-site to deliver the waste to the publicly owned facilities. This compulsory use or monopoly status has been one of the most crucial, and most controversial, measures taken by European governments in their support of advanced hazardous waste management systems. The importance of creating a monopoly for the public facilities can be appreciated by examining the near failure of a similar publicly owned facility that did not have such status in the West German state of North Rhine-Westfalia.

North Rhine-Westfalia offers a useful contrast to its neighboring states, Hessen and Bavaria, in that it has attempted a mixture of competing private and state-financed facilities for hazardous waste-management. North Rhine-Westfalia is the FRG's major hazardous waste-producing state, generating nearly twice as much hazardous waste as the other ten states combined.[62] The highly industrialized and densely populated Ruhr region is located in North Rhine-Westfalia, which administratively is organized into autonomous cities and counties that have instituted a regional structure, the Kommunalverband Ruhrgebiet (KVR), to deal with common environmental problems. KVR has formed a public company, AGR, to manage hazardous waste. Recognizing a need for facilities with higher environmental standards than those offered by private companies in the region, the AGR has constructed a central incineration facility at Herten and has planned collection stations to funnel waste to the facility. This system was fully financed by AGR.[63]

This plan for an independent public system to supplement the private waste-handling sector has not been fully successful. The Herten incineration facility has serious financial problems due mainly to a lack of regulatory direction from the state. The high-priced services offered at Herten cannot compete with less environmentally protective alternatives offered by the private market in North Rhine-Westfalia and other states. For example, a generator of varnish residue can legally choose whether he pays DM 450 per ton for its incineration, DM 200 per ton for its storage at the Herfa-Neurode underground deposit, DM 80 per ton for its disposal in a special waste landfill, or, after mixing it with sand, DM 5 per ton for its disposal in a construction rubble landfill.[64] Even strict standardization and regulation at the state level would not prevent generators from taking advantage of laxer regulations in other states unless an export ban were also imposed.

The crucial difference between the Herten operation and those in the neighboring states of Hessen and Bavaria is the lack of legislation requiring compulsory use of the public facilities. AGR has no legal authority to impose such restrictions, and the state authorities have been reluctant to legislate a public monopoly at the expense of the many smaller handlers. As a result AGR has heavily subsidized incineration of hazardous waste at its Herten facility to ensure that the facility will be used. In 1985 the facility was charging approximately half of the cost of incineration to its customers, with the subsidy coming from charges for burning municipal garbage in huge mass-burn units at the Herten facility.[65]

The state government has not been unresponsive to the AGR's problems. In the summer of 1985 the Minister of Nutrition, Land Planning, and Forestry instructed hazardous waste transporters in North Rhine-Westfalia that unless they agreed to deliver more waste to the Herten facility, the Ministry would consider supporting a change in legislation to require shipment to Herten. The waste transporters agreed to transport a portion of the waste collected to the AGR facility at an increased price, which will be gradually increased over four to five years to a level that allows full cost recovery for Herten.[66]

The experience in North Rhine-Westfalia underscores the importance of the historical conditions in which the relatively successful waste management systems in Hessen, Bavaria, and Denmark were put into place. A desire on the part of public authorities to finance an environmentally comprehensive system of management facilities, combined with the explicit cooperation of waste generators and little political opposition from a private waste disposal industry, was important, though not entirely sufficient. In addition, the regulators needed the legal authority to assure a market for the public facilities, which they do not have in North Rhine-Westfalia. In the absence of standardized regulations guaranteeing the use of the best available technologies for hazardous waste management within a nation or among nations, restricting waste exports is the only way to protect large investments in hazardous waste treatment facilities.

All of the public waste management companies discussed in the first part of this chapter operate more or less as statutory monopolies. The export restrictions of Bavaria and Hessen in the FRG have come under attack both within the FRG and from the outside. The Commission of the European Communities, in keeping with its mandate to promote free trade among member countries, is presently challenging the Bavarian and Hessian export controls in the European Court.[67] This is one of the first cases where a country or region may not be allowed to keep its pollution within its own borders, and shows a direct conflict between the aims of the free market and environmental concerns. Despite this challenge, the West German Parliament passed a Third amendment to the Waste Law in 1985 that requires generators to justify to state authorities why waste cannot be managed in the state or nation before it can be exported.[68]

The FRG is also attempting to make practices more consistent among the states by adopting a regulation specifying a preferred technology for each waste in the waste catalog. This regulation is being developed by a working group composed of state waste officials and experts from the Federal Environmental Agency (see Appendix 3 for excerpts). As seen in the United States prior to the adoption of the 1984 RCRA amendments, consistent regulations on a nationwide basis can impede technological progress in waste management where the regulations lock in inappropriate technologies, such as landfilling and deep-well injection. States attempting to set higher standards still face the export problem in their efforts to encourage superior technologies.

Assuming that there is a need for monopoly status for hazardous waste treatment facilities, other models besides the publicly owned, non-profit companies in the European nations discussed can be envisioned from the policy tree in Figure 5.1. These models (numbers 7–10) are all variations of the public utility model: privately owned, regulated monopolies. The public utility model has a long history and precedent in the United States, providing such public services as electricity, water, and gas. Generally these privately owned firms are profit maximizers, but their prices and other practices are closely regulated by the public authorities.

In many respects, the management of hazardous industrial waste would fit naturally within this model. Hazardous waste management has some aspects of a natural monopoly, such as high capital investment costs, a finite market, and the possibility of single, integrated facilities serving whole market areas. Hazardous waste management also has attributes of public necessity, due to the dangers of improper management. The wealth of experience and tradition with regulated public monopolies in many countries would greatly facilitate acceptance of this type of organization for hazardous waste management where publicly owned facilities would be disfavored.

Integrated versus Specialized Facilities: The Third Branch

An important choice facing many governments is whether to encourage the construction of centralized, integrated hazardous waste facilities serving a diverse clientele of hazardous waste generators or whether, instead, to encourage a greater number of smaller-scale, specialized facilities. The distinction between the two choices is not entirely clear-cut, however, since specialized facilities may serve a large number of generators and regional systems may not be integrated but consist of several more specialized facilities.

An important differentiating characteristic of an integrated facility concerns the point in the waste life cycle where the decision is made about how the waste will be managed—whether incinerated at high temperatures, treated, disposed of on land, mixed with other waste, or otherwise processed. Where the facilities are integrated, with all or most of the possible handling techniques under one management, this decision is logically made internally. If the facilities are specialized, the choice will logically be made by the generator or transporter in deciding where to ship the waste.

This distinction becomes important when considering the extreme complexity of a waste-handling system made up of thousands of autonomous generators and transporters with sometimes limited knowledge of the properties of their waste. The most important element of control may be simplification, and an integrated facility, where the only rule is that the generators deliver their waste, reduces the system's complexity. An integrated waste management facility, by condensing the hazardous waste life cycle into a relatively comprehensive regulatory management organization, reduces significantly the decision points and subsequently enhances the possibilities for effective control. The authorities can better oversee the operation of a small number of facilities, which operate with a staff large enough to include environmental experts, than a multitude of smaller operations.

While centralized facilities may be easier to monitor, they have some drawbacks. One important problem concerns the distances that transporters must haul wastes, with the consequent increase in transportation and risks (the relative additional costs are considered to be fairly insignificant).[69]

A second disadvantage regarding larger facilities is the potential dangers of concentrating many different types of wastes in large quantities. The bureaucracy of the facility becomes more meshed in administrative details as waste traffic increases, especially at that point where there is no longer an identifiable and stable relationship between the handler and his clientele. In other words, small specialized firms have the advantage of a relatively constant stream of wastes that they are experienced in handling and for which they are generally clear about the composition.

Allocation of the Costs: The Fourth Branch

At the core of any hazardous waste management system is the question of who pays the costs. Despite directives adopted by the Organization for Economic

Cooperation and Development to encourage member nations to require industrial polluters to pay the full societal costs of their pollution,[70] there is a strong tradition in Europe of the government picking up the tab for environmental protection. This is in contrast to the United States, where governments federal and state hesitate to aid industry directly.

In the FRG, which has a strong laissez faire ideology but also a long tradition of government/industry cooperation, there was early reluctance on the part of the Hessian and Bavarian authorities to provide public subsidies to hazardous waste facilities. The early intention of HIM, ZVSMM, and GSB was to operate as non-profit enterprises but to charge industry the full costs of waste management. The incineration and treatment of waste, however, proved so costly in comparison with land disposal that it was politically difficult to require the firms in these two states to pay significantly higher disposal costs than in the rest of the FRG. The subsidies on the part of the public authorities were not, however, in the spirit of allowing industry a "free ride," since prices remain relatively high in these two states.

Opponents of government financial support are concerned about the equity of passing on the costs of hazardous waste management to the general public as well as the resulting price distortions that misplace the incentives for reducing wastes. Indeed, general agreement could probably be reached on the goal that industry should pay the full costs of the management of its hazardous waste. But passing high treatment costs on to the generators, which has the very positive environmental effect of reducing the generation of wastes, creates a dilemma for the treatment facility investors by reducing the waste market and a dilemma for enforcement agents by encouraging generators to dispose of their waste illegally.

Most of the facilities discussed above are troubled with overcapacity. Predictions of waste generation that formed the basis for construction of the plants overestimated the quantities of waste generated. Although stricter enforcement of hazardous waste regulations has resulted in more waste being brought to the facilities, waste reduction, due to higher prices for management, and slow economic growth have combined to counteract any increase in the waste generation rate. The shortage of waste delivered to the facilities is further aggravated by the fact that the waste delivered is becoming "dirtier" as easier to recycle components are separated out.

From the waste reduction perspective, a measure of success of a high-priced treatment facility is how quickly it manages to put itself out of business. But this strategy presents an obvious dilemma to public or private facility owners and underscores the conflicting institutional objectives. A pricing or choice of technology policy that encourages the environmental objective of reducing hazardous waste thus conflicts with the economic objective of maintaining a reasonable return on the large capital investment in waste facilities. This creates a fundamental dilemma between those institutional interests charged with reducing environmental risks and those charged with creating and sustaining the treatment infrastructure.

The German States of Hessen and Bavaria, in the face of decreasing supplies of hazardous waste, have taken the following stopgap measures to sustain the treatment infrastructure:

1. Expansion of the universe of hazardous waste by adding wastes to the list

2. Restriction of the amount of waste that is exported by requiring all waste, unless officially exempted, to be brought to the public facilities

3. Increasing the amount of waste delivered by restricting on-site disposal

4. Importing waste from other states and countries

These measures, apparently motivated both by the need to maintain a supply of wastes to the state facilities and by a genuine distrust of the private sector in handling waste, represent a compromise between the conflicting objectives of reducing waste and sustaining the capital-intensive facilities. Yet they will not suffice in the long run as generators find more ways to reduce costs by reducing waste generation. This dilemma might eventually resolve itself when the facilities become obsolete. Meanwhile, Bavaria plans to phase out its subsidies to the hazardous waste management companies.

The opposing institutional objectives prescribed by sustaining highly qualified facilities and at the same time reducing hazardous waste generation are even further complicated, since the high treatment prices that encourage waste reduction are at the same time an incentive for illegal disposal. The success of an environmentally adequate but costly management system depends ultimately on whether it encourages more legitimate recycling and reduction measures while controlling illegitimate dumping, burning, or exporting.

With thousands of firms producing and transporting hazardous waste and numerous opportunities to evade the regulatory system, full control is close to impossible, especially considering the limited funds available in government budgets for this purpose. A serious disparity then arises between the legislative efforts to control hazardous wastes and the implementation of this legislation, which has been referred to as the "enforcement gap"[71] or "executive deficit."[72]

In both the United States and European nations the difficulties are so great in policing the complex system of waste management that regulators repeatedly stress the importance of information dissemination and moral persuasion in encouraging industry to comply.[73] Indeed, figures show that the cost of control may be a significant portion (20–40 percent) of overall disposal costs.[74] With the limited financial resources for enforcement, the regulators can at most inspect and monitor the large generators. In the United States, the largest 5 percent of hazardous waste generators produce approximately 98 percent of the waste.[75] But this may be where efforts are least needed since the large firms are generally highly sensitive to hazardous waste problems and can generally be trusted to comply with regulations.[76]

Do subsidized, publicly owned facilities help narrow the enforcement gap?

The evidence is unclear. There are opposing opinions on the extent of illegal disposal practices in Bavaria and Hessen, for instance. Officials at the state and local levels have expressed the opinion that illegal disposal practices are no longer a significant concern; the problem of midnight dumping, they claim, is a problem of the past.[77] But officials of the West German Federal Environmental Agency (UBA) are not as sanguine about compliance in the German states, estimating the extent of illegal practices to be anywhere from 10 to 90 percent, or, at best, highly uncertain.[78] Still the general perception on the part of most officials is that the relatively high management prices in Hessen and Bavaria have not increased the amount of illegal dumping. The extent of these practices cannot be readily estimated, but many experts feel that illegal disposal is less of a problem in the FRG, generally, than in other EEC countries.[79]

The more serious problem in the FRG, it seems, is the transport of hazardous waste across state or national borders to less qualified facilities. In theory, a waste management firm in other West German states should not accept waste that is transported without permission from Hessen or Bavaria. In practice, officials state that it is not difficult for generators to find out-of-state firms willing to take their wastes. There is also a lucrative waste traffic to East Germany and other countries.

In Denmark officials believe they have captured most of the chemical waste generated by industry in the system of municipally owned transfer stations and the Kommunekemi facility. But as noted earlier, transfer stations that institute their own collection services usually note an increase in the amount of waste received. The national government also has not monitored until recently the exemptions granted to firms from the delivery requirement. Furthermore, in 1984 a Danish firm was caught disposing of chemical waste in a municipal dump and was the focus of a very public enforcement action. Following this incident, there was a significant increase in the amount of waste received by Kommunekemi, possibly indicating that several other firms had been using illegal practices.[80]

LESSONS FOR AMERICA

Any country intent upon shifting from land disposal to more reliable treatment and incineration methods must choose and promote a physical and economic infrastructure that secures this transition. The choices in this process are difficult since they involve complex trade offs among the sometimes conflicting goals of a waste management system: (1) the promotion of environmentally sound treatment and disposal practices; (2) the support of waste reduction; (3) the assurance of full compliance with the system; and (4) the allocation of the full costs of hazardous waste management to the generators. None of the systems examined in this chapter has been completely successful in fulfilling all four of these objectives.

The European systems examined here have placed a high priority on the first goal by constructing relatively expensive treatment and incineration facilities with relatively little direct land disposal. But this often has occurred at the expense of the fourth goal, since the capital-intensive facilities have been built with a steadily rising investment in public funds. The public authorities have assured a market for these high-cost systems by requiring all waste shipped off-site to be delivered to these facilities, thereby creating a public monopoly. The European models also expose the tradeoff between the third and fourth goals. A significant portion of the capital costs of the facilities has been borne by the taxpayers as a way of lowering prices to prevent widespread illicit disposal and export of hazardous waste.

This public monopoly model stands in sharp contrast to the private competitive models found in German states other than Hessen and Bavaria and to the United States, where scattered facilities are usually owned and operated by private entrepreneurs with no direct government subsidization. The United States, with its emphasis on regulatory controls and generator liability, has placed high priority on the fourth management goal but with some sacrifice in the first goal, as shown by the large percentage of wastes that are disposed on land.

An important conclusion of this chapter is that the private competitive model found in the United States will have difficulty fulfilling the first goal by providing the necessary infrastructure for environmentally sound treatment of the majority of regulated waste, even if the government comprehensively requires this type of management in its regulations. One reason for this difficulty is the inevitable decline in the market for higher-cost methods resulting from the new emphasis on waste reduction and recycling and from export of waste to other states and countries with less stringent regulations. Another is that there is no guarantee that sufficient treatment capacity will be provided by the private sector in a timely, efficient manner for all waste needing treatment. Export controls and partial government subsidization may be necessary to guarantee that capacity and as a means of discouraging the illicit disposal of waste that further aggravates the declining waste market.

Hessen, Bavaria, Denmark, Sweden, and Finland responded to the failure of the private sector by sliding toward public ownership and financial support of large, integrated waste management facilities. Those governments have thus "absorbed" the risks of hazardous waste management and the public has to some extent "absorbed" the costs. The success of these models must be viewed in the historical and political context in which they emerged and operate. The Hessian and Bavarian systems for instance, were set up with little competition from the almost non-existent private waste-handling sector. The financial failure of early private initiatives, combined with a pervasive mistrust on the part of state officials of private entrepreneurs for handling hazardous waste, led to government intervention with the full support of waste generators. Direct government subsidies were also quite consistent with the European tradition of public support for environmental protection.

The restrictions on exports of waste and the compulsory use of central treatment facilities are also critical features of the European story, as illustrated by the near failure of the Herten incineration facility in North Rhine-Westfalia, which did not have an assured market. Dedicated and environmentally concerned public managers were also an important prerequisite for successful management systems.

The apparent success of the systems examined, therefore, hinged on several factors: (1) the absence of a significant private waste management industry to resist increasing public support of the treatment facilities; (2) the ability of the government to restrict exports and on-site management and to legislate compulsory use; (3) the traditional public acceptance of taxpayer support of environmental measures; and (4) the presence of environmentally committed civil servants to manage the facilities. Where one or more of these conditions is lacking, this public monopoly model may not be easily adopted or particularly effective.

In the United States any attempts at creating a public monopoly would likely meet strong resistance from the increasingly powerful private waste management sector, and restricting exports to other states or regions within the nation might be deemed unconstitutional as inconsistent with the commerce clause protecting interstate commerce.[81] American industry has traditionally been held to be responsible for environmental costs, and the subsidization of facilities with public funds may also meet strong political obstacles. For these reasons, direct responsibility for, or "absorption" of the risks of hazardous waste management by federal, state, or local governments may not be feasible or desirable in American political culture.

Public or quasi-public integrated hazardous waste management facilities have been considered in the United States. As early as 1974 the Environmental Protection Agency prepared a report to the U.S. Congress that advocated regional, centralized processing facilities for hazardous waste.[82] No suggestion was made that the government might actually build or operate these facilities, but the report did suggest that the government provide assistance by creating a franchise system with territorial limits, stopping short of granting monopoly status, however.

Attempts at creating large integrated treatment facilities in the United States have largely failed, such as the $100 million integrated treatment facility IT Corporation proposed for Louisiana. An attempt at creating a publicly owned integrated facility by the Gulf Coast Waste Disposal Authority in Texas also failed but has recently been revived. One of the few integrated facilities to be conceived as such in recent years is the state-sponsored facility under construction in Arizona. Arizona is providing the land and has selected a contractor who will construct and operate the facility according to state specifications. This example approaches the public utility model, but without guaranteed waste flow control and monopoly status.

The 1984 RCRA amendments if applied consistently, uniformly, and in accordance with congressional intent, will eventually help create a market for alternatives to land disposal and will resolve the problem of interstate shipments of waste to states allowing less expensive disposal technologies. The sole reliance on regulatory controls, even with this ambitious legislation, will not guarantee that the needed hazardous waste management infrastructure will be set up in an economically and environmentally sound way. Full reliance on private enterprise has serious drawbacks, as pointed out throughout this book.

The generally positive experiences with the public monopoly model in Europe suggest distinct advantages if Americans depart from the notion that a sound hazardous waste management system will evolve only as a product of intense regulatory struggle between private enterprise and government regulators. Safe hazardous waste management is a public necessity and requires greater and more consistent governmental involvement. This does not and cannot mean the wholesale transfer of the European model to the United States, although important elements of the system may be transferrable.

One branch of the policy tree, the public utility model, involves the public planning and selection of the most environmentally sound technologies which are then constructed and operated by private companies with strict public supervision and accountability. State and local governments might grant territorial monopolies or consider appropriate subsidies. The public utility model may offer the greatest promise of balancing the four goals of sound hazardous waste management within American political culture. This approach deserves further research as the United States moves beyond land disposal.

NOTES

1. U.S. Congress, Office of Technology Assessment, *Technologies and Management Strategies for Hazardous Waste Control,* (Washington, D.C., 1983), p. 8.

2. S. Kent, Stoddard, Gary A. Davis, and Harry M. Freeman, *Alternatives to the Land Disposal of Hazardous Wastes: An Assessment for California,* California Governor's Office of Appropriate Technology (Sacramento, 1981); U.S. Congress, *Technologies and Management Strategies for Hazardous Waste Control;* M. Prurot (Rapporteur), "Report on the Treatment of Toxic and Dangerous Substances by the European Community and Its Member States," *European Community,* April 9, 1984, pp. 8–9.

3. L. S. Brasser, Acting Deputy Director of the Netherlands Organization for Applied Scientific Research, personal communication, August 12, 1982.

4. Department of the Environment, *Cooperative Programme of Research on the Behavior of Hazardous Wastes in Landfill Sites, Final Report of the Policy Review Committee* (London: HMSO, 1978).

5. A. Szelinski and C. Wels, FRG Federal Environmental Agency, personal communication, West Berlin, November 17, 1983.

6. 3rd Amendment to the Waste Law, West German Parliament, February 7, 1985.

7. Helmut Schnurer, Leader of the Working Group on Waste Management, FRG Ministry of the Interior, personal communication, July 1985.

8. Carl Otto, Zubiller, Hessian Ministry of the Environment, personal communication, July 1985.

9. B. Furmaier, "Organisation und Stand der Sondermüllbeseitigung in Bayern," unpublished paper, October 1983.

10. Ibid.

11. Bavarian Environmental Protection Agency, *Sondermüll Statistic Bayern*, (Munich, 1983).

12. Ibid.

13. H.-G. Ruckel, "Zweckverband Sondermüllplatze Mittelfranken," unpublished paper, Schwabach, FRG, 1983.

14. Hans Ruckel and Norbert Amsoneit, ZVSMM, personal communication, Schwabach, FRG July 8, 1985.

15. Ibid.

16. ZVSMM Price List, January 1, 1985.

17. Hans Ruckel and Norbert Amsoneit, ZVSMM, personal communication, Schwabach, FRG, July 8, 1985.

18. Franz Defregger, "The Bavarian Hazardous Waste System, Illustrated by the Ebenhausen Treatment Plant and Gallenbach Landfill Site," in J. Lehman, ed., *Hazardous Waste Disposal* (New York: Plenum Press, 1983).

19. Ranier Meixlsperger, Bavarian Environmental Protection Agency, personal communication, Ebenhausen, FRG July 3, 1983.

20. Franz Defregger, Director, Waste Management Division, Bavarian Ministry for Regional Development and Environmental Affairs, personal communication, Munich, July 9, 1985.

21. GSB Price List, January 1, 1985.

22. Hessian Ministry of the Environment, *Sonderabfallbeseitigung*, Argumente in der Umweltdiskussion, June 1983.

23. Hessian Ministry of the Environment, *Waste Catalog* (Wiesbaden, FRG, November 1981).

24. Ibid.

25. Carl Otto Zubiller, Hessian Ministry of the Environment, personal communications, Wiesbaden, FRG, July 1983, July 1985.

26. Hans Christoph Boppel, member of Hessian State Parliament, Green party, personal communication, Wiesbaden, FRG, July 11, 1985.

27. Hessian Ministry of the Environment, *Sonderabfallbeseitigung;* Argumente in der Umweltdiskussion, June 1983.

28. Carl Otto Zubiller, Hessian Ministry of the Environment, personal communication, Wiesbaden, FRG July 11, 1985.

29. Ibid.

30. F. Vahrenholt, "Zur Situation der Sonderabfallbeseitigung in Hessen," in Konzepte und Methoden der Sonderabfallbeseitigung, Protokoll des Weisbadener Sonderabfall-Seminars ("The Status of Hazardous Waste Management in Hessen," in *Concepts and Methods of Hazardous Waste Management*, Wiesbaden Special Waste Seminars, 1984), p.6.

31. Carl Otto Zubiller, Hessian Ministry of the Environment, personal communication, Wiesbaden, July 11, 1985.

32. Ibid.

33. HIM Price List, January 1, 1985.

34. Vahrenholt, "The Status of Hazardous Waste Management in Hessen," p.6.

35. Mogens Palmark, "The Danish System," paper presented at the 2nd International Symposium on Operating European Centralized Hazardous (Chemical) Waste Management Facilities, Odense, Denmark, September 1984.

36. Ibid.

37. ChemControl A/S, *The Danish System,* (Copenhagen, Denmark, 1980).

38. Palmark, "The Danish System."

39. Per Riemann, Plant Manager, Kommunekemi, personal communication, Nyborg, Denmark, June 1983).

40. Kristen Warnoe, "The Controlled Landfill," paper presented at the 1st International Symposium on Operating European Centralized Hazardous (Chemical) Waste Management Facilities, Odense, Denmark, September 1982.

41. Handling Prices, Kommunekemi, January 1, 1985.

42. Palmark, "The Danish System."

43. Jorgen Lauridsen, "The Legislation Regarding Oil and Chemical Waste in Denmark," paper presented at the 1st International Symposium on Operating European Centralized Hazardous (Chemical) Waste Management Facilities, Odense, Denmark, September 1982.

44. Jorgen, Lauridsen, Danish Environmental Protection Agency, personal communication, Copenhagen, Denmark, June 1983.

45. Lauridsen, "Legislation."

46. Jorgen Lauridson, Danish Environmental Protection Agency, personal communication, Copenhagen, Denmark, June 1983.

47. Lauridsen, "Legislation."

48. Jens Kampmann, "Benefits from and the Problems Experienced with the Danish Hazardous Waste Management System," paper presented at the 2nd International Symposium on Operating European Centralized Hazardous (Chemical) Waste Management Facilities, Odense, Denmark, September 1984.

49. Jorgen Lauridsen, Danish Environmental Protection Agency, personal communication, Copenhagen, Denmark, June 1983.

50. Ibid.

51. Karl-Erik Kulander and Peter Solyom, Swedish Water and Air Pollution Research Institute, personal communication, Stockholm, July 22, 1985.

52. Ibid.

53. Lars Llung, Manager, SAKAB, personal communication, Norrtorp, Sweden, July 23, 1985.

54. "This is SAKAB," SAKAB, Norrtorp, Sweden, 1984.

55. Lars Llung, Manager, SAKAB, personal communication, Norrtorp, Sweden, July 23, 1985.

56. F. Aarnio, "The Finnish System," paper presented at the 1st International Symposium on Operating European Centralized Hazardous (Chemical) Waste Management Facilities, Odense, Denmark, September 1982.

57. Ibid.

58. Ibid.

59. In contrast, the West German Council of Economic Advisors advised against substitution of publicly owned waste management facilities for privately owned management. "Summary of the Environmental Report," FRG Council of Economic Advisors, February 1978.

60. B. Furmaier, Bavarian Environmental Protection Agency, personal communication, Munich, March 30, 1983; Hans Ruckel, Director, ZVSMM, personal communication, Schwabach, FRG, July 8, 1985; Carl Otto, Zubiller, Hessian Ministry of the Environment, personal communication, July 11, 1985.

61. "Alternatives to the Land Disposal of Hazardous Waste: the European Experi-

ence,'' California Foundation on the Environment and the Economy, San Francisco, May 1983.

62. W. Schenkel, "Sonderabfallbeseitigung in BRD," in W. Kemerling, ed., *Sonderabfall und Gewasserschutz,* 19th Seminar of the Austrian Water Pollution Assoc., Gmunden, Austria, March 5–8, 1984.

63. W. J. Dewey, "Disposal of Hazardous Waste in Northrhine-Westfalia," paper presented at the 2nd International Symposium on Operating European Centralized Hazardous (Chemical) Waste Management Facilities, Odense, Denmark, September 1984.

64. Schenkel, "Sonderabfallbeseitigung in BRD."

65. Johann Fiolka, Planner, AGR, personal communication, Essen, FRG, July 12, 1985.

66. Ibid.

67. Ludwig Kramer, Attorney, Commission of the European Communities, personal communication, Vienna, Austria, July 3, 1985.

68. 3rd Amendment to the Waste Law, West German Parliament, February 7, 1985.

69. J. Linnerooth, *The Transportation of Dangerous Goods: International Comparison of Legislation and Implementation,* Draft Report, International Institute of Applied Systems Analysis, Laxenburg, Austria, 1985.

70. Organization for Economic Cooperation and Development (OECD), *The Polluter Pays Principle: Definition, Analysis, and Interpretation* (Paris, 1975).

71. B. Wolbeck, "Political Dimensions and Implications of Hazardous Waste Disposal", in J. Lehman, ed., *Hazardous Waste Disposal* (New York: Plenum Press, 1983).

72. R. Spilke, "Germany: A Black Hole in the North Sea for Toxic Wastes", *Ambio* 11, no. 1: 57.

73. Colin Diver, "A Theory of Regulatory Enforcement," *Public Policy* 28 (1980): 257–301.

74. Schenkel, "Sonderabfallbeseitigung in BRD," p.3.

75. S. Senkan and N. Staufer, "What to Do with Hazardous Waste," *Technology Review,* December 1981, p. 34–49.

76. Hans Sutter, FRG Federal Environmental Agency, personal communication, April 8, 1983.

77. B. Furmaier, Bavarian Environmental Protection Agency, personal communication, Munich, March 30, 1983; Hans Ruckel, Director, ZVSMM, personal communication, Schwabach, FRG, July 8, 1985; Carl Otto, Zubiller, Hessian Ministry of the Environment, personal communication, July 1985.

78. A. Szelinski and C. Nels, FRG, Federal Environmental Agency, personal communication, November 17, 1983.

79. B. Wolbeck, FRG Federal Ministry of the Interior, personal communication, Bonn, April 24, 1983.

80. Kampmann, "Benefits and Problems."

81. The commerce clause of the U.S. Constitution states that the Congress shall have power to "regulate commerce . . . among the several states." U.S. Constitution, article I, section 8, clause 3. In the U.S. Supreme Court case of *Philadelphia v. New Jersey,* the court struck down New Jersey's prohibition on import of out-of-state garbage as contrary to the commerce clause. 437 U.S. 617, 628 (1978).

82. U.S. Environmental Protection Agency, "Report to Congress on Hazardous Waste Disposal," Washington, D.C., 1974.

6

Making Waste Recoverable: European Waste Oil Programs

LEE BRECKENRIDGE

INTRODUCTION

Throughout the United States, bankrupt used oil processing facilities rank high on the lists of hazardous waste sites needing government-funded cleanup. The unscrupulous practice of mixing hazardous wastes with used oil destined for sale as a fuel or a dust suppressant on roads has resulted in serious incidents of water, soil, and air pollution. Public attention has focused in recent years on a major policy issue: How can the government encourage the recovery and reuse of waste oil while preventing the harmful environmental impacts of waste oil misman-agement?

Waste oils are recyclable but potentially dangerous wastes that pose a maze of regulatory and enforcement problems. These materials are sometimes carefully and safely recovered as a valuable resource, but sometimes indiscriminately dumped, carelessly managed, or reused in a contaminated form with disastrous consequences. In examining waste oil management practices, we exemplify problems that are encountered in the management of many different kinds of potentially recyclable chemical wastes.

In the United States waste oil has been exempt from national hazardous waste management regulations, and the federal government is only beginning to take steps to develop a regulatory framework for controlling the environmental hazards

The author's study of European waste oil programs was made possible by a grant from the German Marshall Fund. The views expressed in this chapter are those of the author as an individual and do not represent any official position of the Attorney General of the Commonwealth of Massachusetts.

This chapter was edited by Gary A. Davis and Bruce Piasecki.

of waste oil management practices.[1] The newness of the federal regulatory efforts in the United States is in contrast to the longstanding programs of several European countries.

In Europe comprehensive waste oil management programs predated more generalized hazardous waste legislation and regulations. The European Communities' 1975 Directive on the Disposal of Waste Oils embodied principles already developed in the legislation of various member countries.[2] Its terms not only reflected a goal of controlling the environmental impacts of waste oil management, but also revealed a deep-seated concern for reducing European dependence on foreign oil imports by recycling existing oil resources. A decade after the original issuance of the Waste Oil Directive, Europeans have now addressed the need for a second generation of legislation. As the United States embarks upon its initial regulatory effort, it is telling to look at the European experience.

PROBLEMS INHERENT IN WASTE OIL MANAGEMENT

Two concerns motivate government control of waste oil. First, waste oil is a valuable national resource. Recycled oil can compete with and substitute for virgin oil in the fuel oil and lubricants market. Through greater recycling of oil, a nation can reduce its dependence on foreign oil imports, improve its balance of payments, and become less vulnerable in an energy shortage. Moreover, by ensuring the existence of a competitive recycling industry it can avoid monopolistic pricing of virgin oil products. A system to collect and reuse waste oil thus serves a nation's economic interests in a primary way.

Second, the mismanagement of waste oil presents significant environmental risks. All available methods of handling waste oil have the potential for causing serious pollution. If dumped in waterways, even small quantities of oil deplete the available oxygen and poison aquatic organisms. When discharged into a sewer system, oil can disrupt the biological treatment process at the sewage treatment plant. If waste oil is burned, there is a risk that contaminants (such as lead or chlorinated organic compounds) will result in toxic air emissions. If treated to remove contaminants, whether by simple methods or with sophisticated rerefining technologies, the processes inevitably result in by-products and residues that must be disposed of through burning, dumping, or discharge. The environmental impacts of waste oil management are consequently an important government concern.[3]

Establishing a workable waste oil program is a complicated task, however. A 1983 study prepared for the Commission of the European Communities estimated the total quantity of waste oils generated each year in the member states at 50 percent of total lubricant use, or 2.0 to 2.5 million tons.[4] In the United States as well, approximately one-half of all lubricants sold each year becomes waste oil, while the rest is consumed during use.[5] The difficulties of controlling waste oil management stem from the multiple types and sources of waste oil, the large amounts of waste oil generated each year, the wide variety of possible contaminants, the multitude of handling or disposal methods, and the complexity of the economic factors affecting waste oil management practices.

In addition to crankcase oils from cars and trucks, there are hundreds of other lubricating oils with thousands of industrial applications, such as hydraulic oils, metal-working oils, turbine oils, electrical oils, gear oils, and railroad and marine oils. To this list of "used oils" one can add other oils that become waste through spillage or contamination without being used, such as the oil in ballast water from ships.[6] The waste oils that are generated by all of these numerous sources vary greatly in their characteristics and composition, further complicating the task of formulating a regulatory control system.

Waste oils may contain many different kinds of contaminants. First, many varieties of additives are now introduced into lubricating oils to enhance their suitability for particular purposes. Second, metals and toxic organic compounds often get into waste oils as a result of their use and handling. One noteworthy contaminant in this category is lead, which typically comes from leaded gasoline that contaminates the oil in automobile engines during use. Since small amounts of lead can have serious toxic effects on human health when inhaled or ingested, the air emissions of lead during waste oil burning are a cause for particular concern. Many other contaminants such as polychlorinated biphenyls (PCBs) and chlorinated solvents are added to waste oils through deliberate mixing or careless practices.[7]

Effective government control is further complicated by the large number of handlers and facilities for reprocessing or disposal. Waste oil is often simply dumped on the ground or into sewers and waterways. In the United States, though not in Europe, it has been widely used for spraying on roads as a means of dust suppression. Waste oil may also be burned "as is" or after simple processing to separate oil and water fractions and to remove solids. Finally, waste oils may be regenerated for reuse as lubricants. While some used light industrial oils can be returned to use as lubricants through relatively simple processes, most waste oil requires separation into component fractions and removal of impurities through rerefining.[8]

THE ECONOMICS OF WASTE OIL MANAGEMENT

Economic factors further complicate this picture. Competition from virgin oil products creates a ceiling on the sale price of the recycled products. When the costs of collecting and processing waste oil exceed the potential resale price, there is no economic incentive for producers, collectors, fuel processors, or rerefiners to participate in the recovery of waste oil. As a result, waste oil that is generated in small quantities or at remote locations tends not to enter the collection system. The "do-it-yourself" motorist, for example, has only a few quarts of used motor oil at any one time. The isolated service station in a rural area may be far from any collection or processing center. The willingness to deal with small quantities may change over time, however, as prices in the fuel and lubricants markets change. Economic factors thus encourage the unregulated dumping and burning of uncollected waste oil, but identifying which waste oils will be improperly managed in the absence of government control is difficult, since the markets for waste oils change frequently with the vagaries of economic conditions.

Market pressures often promote inexpert burning of waste oil that enters the collection system. In the 1970s fuel oil prices rose dramatically without a corresponding rise in the price of lubricants. Waste oil can be burned with little or no processing, while rerefining of lubricants is a sophisticated and increasingly expensive process. Environmental control programs in some countries, including the United States, have compounded the price disparities by imposing restrictions on the disposal of acid sludges from rerefining while failing to control the air emissions from waste oil burning. Market prices and the relative costs of processing consequently have strongly favored the burning of waste oils, diverting waste oil feedstock away from lubricant recovery at rerefineries.

These economic pressures raise important policy questions. Should a government waste oil program favor rerefining over burning? From the standpoint of preserving national resources, rerefining is preferable. From the environmental standpoint, rerefining and burning of waste oil both present risks, but burning raises the threat of toxic air emissions in widely dispersed locations in a way that the centralized rerefining of lubricants does not. Different European governments have made different judgments about whether to favor one mode of recycling over another. The governments that are discussed below share the view, however, that the goals of resource conservation and environmental protection both depend upon strong measures to ensure the collection of waste oil and to control waste oil management practices.

THE WASTE OIL DIRECTIVE OF 1975: A MAJOR FIRST STEP

The Directive on the Disposal of Waste Oils adopted by the Council of the European Communities on June 16, 1975, and amended on December 22, 1986, is an unusual piece of legislation. First, unlike most national environmental laws, the directive is not limited to a particular type of environmental problem such as air pollution or water pollution. Rather, it requires member states to address the environmental problems of waste oils from an integrated perspective. Second, instead of relying exclusively on prohibitions and restrictions, the directive also envisions government programs that affirmatively recover waste oils and channel them toward beneficial uses.

The directive embodied many aspects of national laws and regulations that individual member states had already proposed or promulgated. Indeed, a major purpose of the directive was to enhance the operation of a common waste oil market by eliminating inconsistencies among national laws governing waste oil management. Some member states had already developed regulatory controls and systems of financial incentives to influence the handling and use of waste oil prior to 1975. The Federal Republic of Germany, in particular, had instituted a comprehensive and effective system of waste oil collection under a waste oil statute enacted in 1968, and parts of the European Communities' Waste Oil Directive drew heavily on that German legislation.

The directive mandates control of waste oil handling from the moment of generation through recycling or disposal.[9] Each member state must ensure safe collection and disposal of waste oil,[10] and it must maximize recycling of waste oil through regeneration of lubricants or combustion as fuel.[11] The waste oil management mechanisms envisioned by the directive are a combination of prohibitions, mandatory duties, and financial incentives. While some of the mechanisms are familiar to Americans, others entail a degree of government involvement unusual in the United States.

The directive requires member states to establish technology-based pollution limitations, permit requirements, and record-keeping duties for the purpose of preventing environmentally unsafe practices. Each member state must have legislation to prohibit water pollution and soil pollution, and to prevent uncontrolled air pollution and discharge of residues from waste oil processing.[12] There must be an inspection and permitting system for facilities that "dispose" of waste oils (including recycling enterprises).[13] Producers, collectors, and disposers of waste oil must maintain records on the waste oils handled and provide information to government authorities on request.[14]

The required pollution control measures resemble environmental regulations in the United States. The U.S. Environmental Protection Agency imposes similar licensing requirements and pollution limits in regulating discharges to waterways under the Clean Water Act, and in controlling air emissions under the Clean Air Act. The obligations of producers, collectors, and disposers to keep records and make reports about shipments resemble United States regulations governing hazardous waste collection, storage, treatment, and disposal under the Resource Conservation and Recovery Act, but as noted earlier, the United States has not extended such requirements to waste oil management.

Other requirements of the directive go well beyond the measures characteristic of U.S. environmental legislation. If member states cannot accomplish the goals of environmental protection and resource recovery through the regulatory measures described above, they must take affirmative steps to establish collection and disposal services.[15] This obligation implicitly includes providing financing as necessary, if the operation of these services is unprofitable. The directive states that national governments may contract with waste oil collection and disposal companies, requiring them to accept all waste oils within zones assigned by the government, and in return guaranteeing these companies a reasonable profit through government payments.[16] The government funds to make these payments may be raised through a tax on lubricants that become waste oils or on waste oils themselves.[17]

The directive suggests rather than mandates the use of such a financing system.

It does require, however, that any form of government support comply with the "polluter pays" principle and avoid significant impacts on free trade.[18] A fund financed by a tax on lubricants meets these criteria because the "polluter," the user of oil who generates waste oil, pays the government's cost of ensuring proper disposal as part of the price of the original product.

These portions of the directive have no immediate counterpart in the United States. Although the funding mechanism bears some resemblance to the chemical feedstock tax that finances the "Superfund,"[19] the closest analogies to the agreements between government and industry are found in the regulation of enterprises that Americans view as public services: garbage collection, for example, and regulated public utilities such as gas and electric companies. In those situations the state or local government departs from a mere watchdog role in order to actively ensure that the public receives services. Through contractual arrangements or rate-making proceedings governments require designated enterprises to provide certain services in assigned zones, granting them in return a monopoly and the right to operate and to receive financial benefits that are adjusted periodically to provide a "reasonable" rate of return.

At the time of adoption the Waste Oil Directive was widely praised as the first application of a broader waste management policy in the European Communities. Citing the close link between the economic benefits of resource recovery and the environmental benefits of improved collection and recycling, a spokesman from the Commission of European Communities in 1976 called the Waste Oil Directive a model for subsequent Community-wide legislation on other recyclable waste. He predicted that under the national systems of waste oil management envisioned by the directive, rerefined lubricants could come to represent 25–35 percent of the total lubricants market in the European Communities.[20]

Ten years later, however, not all of the member states had enacted specific waste oil legislation. Those countries that had enacted legislation had taken divergent approaches. The member states remained especially divided over the use of subsidies, incentives, and other affirmative governmental measures to direct waste oil reuse. Germany and France, for instance, had established government-sponsored waste oil collection and recycling systems that granted significant financial benefits to the enterprises selected, while Great Britain and Belgium openly resisted taking such affirmative measures. The member states also disagreed about the regulation of recycling methods. France, for example, required waste oil to be rerefined if possible, while the Netherlands had no rerefineries at all and required waste oil to be burned.

The countries with the most extensive systems claimed that discrepancies among member states in the interpretation and application of the directive and the resulting differences in waste oil prices undermined the more ambitious programs. Responding to these concerns, the Commission of the European Communities in 1985 proposed amendments aimed at enhancing and controlling waste oil collection, encouraging rerefining of lubricants as the preferred mode of recycling, and limiting air pollution from waste oil burning. A modified version of the Commission's proposal was adopted by the Council of the European Communities on December 22, 1986. The amendments retained and strengthened the basic provisions of the existing Directive, and added a number of important requirements.

The amended Directive redefines "waste oils" to exclude synthetic oils such as PCBs.[21] Member states must ensure that waste oils used as fuel, and waste oils once regenerated, do not contain more than 50 ppm PCBs and do not constitute a "toxic and dangerous waste."[22] Member states must control the disposal of residues from waste oil recycling and they must regulate the emissions from waste oil burning.[23] While the amended Directive did not adopt the Commission's proposal to ban waste oil burning in small plants, it does require larger plants to be held to specific air pollution limits.[24] Under the new provisions, the regeneration of waste oil must be given priority over combustion, if practicable.[25] The amended Directive also imposes a new requirement that waste oil collectors be registered and supervised, possibly through a permit system.

A major purpose of the amendments to the Directive was to require the member states that were lagging in their regulatory efforts to introduce controls on collection and recycling that were similar to those in countries with more elaborate programs. The discussion below addresses two of the more interventionist waste oil management programs in Europe: those in Germany and France.

WEST GERMANY'S SYSTEM: A MUCH-PRAISED MODEL COMES UNDER NEW SCRUTINY

The most studied and applauded waste oil system in Europe belongs to the West Germans. The Federal Republic of Germany adopted legislation in 1968 to encourage recycling of waste oil well before the European Communities adopted the Waste Oil Directive in 1975.[26] The Waste Oil Directive was modeled in some ways on this German legislation, and the German law was subsequently amended to follow closely the terms of the Waste Oil Directive.[27]

The German waste oil program was substantially revised in 1986, however, in the wake of heated debates over the environmental efficacy of the system. Effective November 1, 1986, the waste oil law was superseded by new legislation that integrates the requirements governing waste oil with the general law on waste management. It is also striking that the aspect of the system most lauded for its effectiveness—a program of subsidies funded through a tax on lubricants— is now being phased out. A discussion of these important developments requires an overview of the way that the system has operated in the past.

The two foundation stones of the German system were the government-sponsored collection program and the subsidy of waste oil recycling enterprises. These aspects of the system went beyond simple pollution restrictions and prohibitions by aggressively channeling the nation's waste oil into recycling.

The Collection System

Under the waste oil law the Federal Office for Trade and Industry[28] ensured the collection of waste oil in quantities over 200 liters and arranged for storage facilities for smaller quantities.[29] The agency accomplished the task of waste oil collection by entering into contracts with private companies. Each collector agreed to pick up all waste oil in quantities over 200 liters within a designated zone; in return it received the right to carry on business in all other zones in the country.

Within its designated zone, the collector was required to pick up relatively uncontaminated waste oil (less than 12.5 percent contaminants) without charge to the producer. If the waste oil contained more than 12.5 percent contaminants, the collector could charge the producer a price that was established in advance in the collection agreement with the government.[30] As long as a mixture or emulsion contained at least 4 percent oil, it fell within the terms of the waste oil law.[31] The collection system consequently covered both waste oils that had a significant market value when collected and those that did not.

While the German system fixed the charges for some oils it did not limit the payments that collectors could make to producers for high-quality oil. In practice, collectors often had to pay producers to obtain their waste oil, when waste oil prices rose in parallel with the prices of virgin fuel oil. Competition existed among collectors because any firm that accepted the duties of collecting waste oil within a designated zone was also free to compete with other collectors for the purchase of waste oil in other zones.

The Subsidy System

The Federal Office for Trade and Industry has also administered a system of subsidizing the recycling companies. Firms that recycle waste oil through re-generation of lubricants or by burning with energy recovery have qualified for a subsidy. The payments have derived from a tax on lubricating oils.[32] Manu-facturers and importers have passed the cost of this levy on to consumers, who have thus contributed indirectly to the fund for dealing with the waste oil they produce. In this way, the system has operated on the general premise that the costs of lubricating oils should include the costs of ensuring their subsequent proper disposal, in accordance with the "polluter pays" principle.

Until 1980 the Federal Office for Trade and Industry calculated the subsidy rates by gathering and analyzing extensive statistics from all of the recycling enterprises about their costs. In an effort to avoid supporting inefficient enter-prises, it calculated the subsidy based on the average costs of the enterprises as a group. As a result, the government's support was the same for all enterprises, paid at a periodically adjusted rate for each 100 kilograms of waste oil processed. The costs that formed the basis for the subsidy calculation included not only the costs of collecting and transporting waste oil but also the costs of disposing of waste residues from processing. The program thus covered the full spectrum of environmental management costs. Since the purpose of the subsidy was to ensure the collection and recycling of waste oil that would not otherwise be recycled, the subsidy did not cover the costs of paying waste oil generators for oil that could be profitably recycled without government financial support.

In 1984 the government began a process of phasing out the waste oil subsidy, with the intent of eliminating it entirely by 1990. The subsidy has been established each year at 1 DM less than the previous year's rate.[33] Having fostered the growth of a comprehensive collection and recycling industry, the German government felt that the waste oil management system could be self-maintaining.

The Resource Recovery Record of the German Program

The German waste oil program was immensely successful in encouraging resource recovery and in preventing pollution from uncontrolled dumping and burning. The statistics prepared by the Federal Office for Trade and Industry for 1983 show that authorized collectors picked up 77.4 percent of the waste oil generated in 1983. A campaign to induce individual motorists to participate in the collection system was highly effective. Even "do-it-yourself" motorists who changed their own oil delivered about 85 percent of their used oil to designated collection facilities. With the support of the government-managed fund, an estimated 54.3 percent of all waste oil in the country was delivered to rerefineries for regeneration as lubricants, and only 2.4 percent was burned as fuel. An additional 41.7 percent was recycled without subsidy payments, primarily through burning as fuel in industrial facilities. Of the waste oil generated in 1983 only 0.1 percent was sent to landfills and 1.5 percent was dumped in unknown places.[34]

These figures are in stark contrast to practices in the United States. In the United States in 1983 less than 7 percent of the used oil went to rerefiners for regeneration of lubricants. Generators dumped 20 percent of all used oil—on land, in watercourses, and in sewers—and they sent an additional 9 percent to landfills. "Do-it-yourself" motorists burned, dumped, and disposed of in landfills 86 percent of their used oil, delivering only 14 percent to collection points and service stations. Spraying roads with waste oil for dust suppression accounted for 6 percent of the used oil in 1983. Nearly all of the waste oil that was recycled was burned, not merely in large industrial installations but also in small ill-equipped space heaters and non-industrial boilers without any air pollution equipment.[35]

Despite some minor differences in the way the statistics were calculated, the broad picture is clear.[36] The German program has led the producers of waste oil to turn their waste oil over to centralized recycling facilities through a government-supervised collection system. This has dramatically eliminated the practice of discarding even small quantities of waste oil improperly. While Germans recycle most of their waste oil, Americans discard large quantities without recovering either fuel or lubricants. In short, the Germans have been much more successful in ensuring collection of waste oils and in preventing the environmental consequences of indiscriminate dumping and burning.

The PCB Scare: Contaminants Traveling the Recycling Route

In Germany these successes of the system were acknowledged, and nearly taken for granted. Government and public attention turned to another matter, the failure of the waste oil system to prevent the mixing of toxic materials with the waste oils destined for recycling. While environmental hazards from uncollected waste oil were largely eliminated, the threat of pollution from recycled oil became the subject of intense debate.[37]

The widespread contamination of waste oil resulted from a lack of coordination between the waste oil law and the newer, stricter waste disposal law. The waste oil law, as discussed earlier, contained a very broad definition of waste oils, covering all mixtures with at least 4 percent oil regardless of the nature of the other contaminants. While the law also contained a prohibition against mixing hazardous waste with waste oil, the prohibition was difficult to enforce. The German Waste Disposal Act, meanwhile, explicitly adopted this definition and exempted from its control all materials falling within the scope of the waste oil system.[38] The result of these complementary provisions in the two acts was that many hazardous chemical wastes mixed with waste oil escaped control under the waste disposal law.

Intense public debate and discussion within agencies at all levels of government followed the proposal of amendments to rectify these waste oil contamination problems. The discussions centered almost exclusively on the contamination of waste oils with PCBs. The "PCB scare" was a new development in Germany. In the course of a few years various incidents turned the PCB contamination of waste oil in Germany into a media event. The well-publicized bankruptcy of Pintsch, a waste oil recycling firm, left the government with the task of cleaning up widespread PCB contamination at its facilities. Heating oil marketed in Berlin was found to contain high levels of PCBs. Analyses also showed high PCB levels in waste oils delivered to rerefiners, and such contamination was not removed in the rerefining process but passed through into the regenerated lubricants. Ironically, the country's successful program of encouraging the regeneration of waste oils as lubricants perpetuated the PCB contamination of the past through repeated recycling.

Widely differing proposals were advanced by two working groups formed by the state governments in Germany.[39] The LAGA (Länderarbeitsgemeinschaft für Abfall, or Interstate Working Group on Waste) recommended that a 50 parts per million (ppm) PCB limit should be included in the definition of waste oil in the waste oil law, and that this limit should be lowered to 10 ppm after three years. The 50 ppm limit was borrowed from U.S. regulations, which have drawn the line at this level of contamination between PCB wastes requiring high-temperature incineration and waste oils that may be burned without special controls.[40] The LAGA's recommendations were adopted by the Conference of Environmental Ministers (Umweltministerkonferenz [UMK]), representing each of the West German states.[41]

A counterproposal was initiated in May 1985 by the Interstate Working Group on Air Pollution, the LAI (Länderausschussimmissionschutz). An LAI consultant found dioxins and furans in the emissions from burning waste oil in a space heater. Noting its responsibilities for preventing air pollution, the LAI urged the Conference of Environmental Ministers to rethink the 50 ppm recommendation and to recommend a limit of 0 ppm PCBs and 0 ppm dioxins.[42] Arguing against the LAI proposal, the LAGA subcommittee questioned the testing results, noting that space heaters were not used in Germany for burning waste oil, and that only one analysis for dioxins and furans had been conducted.[43]

Within the German federal government there was considerable concern about the practical effects of imposing the stricter limits proposed by the LAI. With a 50 ppm limit on PCBs, from two-thirds to three-quarters of the waste oil in the country would still fall within the terms of the waste oil law, but the rest would be treated as a hazardous waste.[44] Under the LAI proposal 95 percent or more of the waste oil in the country would no longer fall within the terms of the waste oil law. In effect, the waste oil program would be eliminated.

The rerefining industry supported stricter controls than under existing law to prevent contamination of waste oil, but it opposed the LAI proposal as unworkable. A representative of Maier, the firm that handles 70–80 percent of the waste oil in the country, said that a 0 ppm limit on PCBs in waste oil would require the company to stop operating. He said, however, that a 50 ppm limit was workable and, indeed, that the company was already required to meet this limit by the local authority.[45] Some of the local authorities had already taken steps, without explicit legislative authorization, to exclude contaminated waste oil from the waste oil management system.[46]

The large oil-producing companies were strongly opposed to either of the proposed amendments because of a concern that the establishment of stricter limits on contamination would make waste oil management costs go up by causing many recyclable oily mixtures to be treated as wastes.[47] The interest of the large oil producing companies in Germany was in ensuring a broad waste oil management system that would provide inexpensive collection and disposal to users. Not surprisingly, they supported combustion rather than rerefining of waste oil as a preferred method of recycling.[48]

Effective November 1, 1987, the new German waste law superseded the existing waste oil and waste disposal statutes. The new law put the waste oil management program in Germany on a new legal footing. Unlike earlier legislative proposals, the new law did not simply narrow the scope of the waste oil law to exclude dangerously contaminated oils, but rather integrated the entire

waste oil program into the more general provisions governing waste management.[49] The new law imposes strict requirements on the management of waste
oil, with the goal of preventing contamination and promoting safe recycling.

Producers of waste oil have the obligation to ensure the proper management
of their waste oil.[50] Producers, collectors, and disposers of waste oil are required
to follow a manifest system for tracking shipments, and they must maintain
records for inspection by government authorities.[51] These documentary requirements supersede the less stringent record-keeping requirements of the waste oil
law.

The law provides for the government to issue orders requiring waste oils with
different levels of contamination to be kept separate and to be managed differently.[52] Under the new program, the LAGA proposal has been adopted, and
waste oils contaminated with more than 50 ppm PCBs are treated as hazardous
waste.[53]

Collectors of waste oil must now obtain a license.[54] The government no longer
has a duty to provide for collection of relatively uncontaminated waste oil free
of charge, and no longer needs to enter into contracts with collectors for that
purpose. The contracts between the Federal Office for Trade and Industry and
collectors under the waste oil law remain valid, however, and are deemed equivalent to licenses until December 31, 1989.[55]

Retailers of motor oils also have new duties under the waste law. Persons that
sell motor oils to consumers must provide oil-changing facilities, accept the
return of waste oil without charge (or arrange for a nearby facility to do so),
and notify consumers of their duty to manage waste oil properly.[56]

The program for subsidizing recycling enterprises is being phased out as
previously anticipated. The preexisting provisions governing the lubricants tax
and the waste oil subsidy remain in effect until December 31, 1989.[57]

The legislative changes leave open the difficult question of where contaminated
materials that previously entered the waste oil system will go. West Germany
does not have the capacity to immediately destroy all of the waste oils contaminated with more than 50 ppm PCBs in incinerators that are authorized to burn
PCBs, since there are only four such facilities in Germany at the present time:
Bayer AG, in Leverkusen (an on-site incinerator at a chemical plant); GSB in
Bavaria (a quasi-public waste management firm); HIM in Hessen (also quasi-
public); and BASF in Ludwigshafen (an on-site incinerator at a chemical plant).[58]

There has been widespread discussion of the possibility of allowing waste oil
contaminated with PCBs to be burned in large industrial facilities such as cement
kilns and power plants that are not authorized to handle PCB wastes.[59] Officials
believe that PCB contamination of waste oil is largely a legacy of past practices
in the mining and electrical industries that could be eliminated in the course of
five years through a concerted incineration program.[60] The only other perceived
alternative is long-term storage of contaminated waste oils or export to countries
that do not enforce strict requirements on waste oil burning.[61]

Although there are many differing viewpoints on the best way to deal with the contamination problems in waste oil, there nevertheless has been a growing consensus in Germany about the need to burn such contaminated waste oils. This policy trend has not precisely coincided with contemporaneous efforts in the European Communities to encourage rerefining instead of burning under the amendments to the Waste Oil Directive.[62]

While the PCB contamination of waste oil has received much public attention, the government's elimination of the waste oil subsidy has proceeded quietly without much apparent opposition. This is despite the fact that more waste oil will likely be burned rather than regenerated, without subsidized recycling. The lack of controversy surrounding the elimination of the subsidy is surprising, since the German system of taxing lubricants and subsidizing waste oil recycling is the aspect of the German law that has been widely cited as a model for other countries, and it is the system that provided the basis for important sections of the Waste Oil Directive.

THE FRENCH SYSTEM: WASTE OIL MANAGEMENT AS A PUBLIC UTILITY

France has gone the furthest in Europe in altering the free-market handling of waste oil by establishing a waste oil program that operates essentially as a public service. As in Germany, the French program uses licensing and government support to ensure the collection and recycling of waste oils. Unlike Germany, however, which has given equal priority to rerefining and burning of waste oil, France has long pursued policies that strongly favor the regeneration of lubricants and preclude burning, even in industrial facilities. Another important difference is that while the Germans have granted long-term subsidies, the French have used grants of monopoly rights, rate setting, and temporary subsidies to support the operation of their waste oil program.

The Laws Authorizing Waste Oil Controls

The tradition of waste oil regulation in France began during World War II. Laws enacted in 1940 allowed purchases of new lubricating oils only by persons who could show that previously purchased oils had been recycled as lubricants. During the war the government fixed the prices of used motor oils and insulating oils, allocated used motor oils among authorized recycling facilities, and set the prices of rerefined motor oils.[63]

In 1956, during the Suez crisis, a government order again required used mineral oil lubricants to be rerefined and prohibited all other methods of disposal.[64] Although enforcement of this order languished in subsequent years, it was invoked again during the energy crisis of the 1970s. Administrative guidelines in 1976 expressed the goal of rerefining all used oils, to the extent possible, as lubricants.[65] France's recycling policy is now definitively expressed in an energy conservation law that provides: ''Mineral and synthetic oils which, after use, are no longer suitable without treatment for the use for which they were intended as new oils, . . . may only be used, if the quality of the oils permits, for regeneration or as industrial fuel. This second use may only be authorized in approved facilities and if the needs of the regeneration industry have been preferentially satisfied.''[66]

The current decrees establishing France's waste oil management system are a specialized application of one of France's two major waste laws, the law of July 15, 1975, governing waste disposal and resource recovery, which gives the government broad powers to ensure recycling.[67] The law authorizes the tracking of shipments of wastes[68] and the mandatory recycling of wastes in authorized facilities.[69] The law even allows the government to interfere in manufacturing processes and to adopt plans allocating waste materials among recycling facilities for the purpose of optimizing resource recovery.[70] In order to encourage these resource recovery projects, the 1975 law created a quasi-public agency, ANRED (Agence Nationale pour la Récupération et l'Elimination des Déchets, or National Agency for the Recovery and Elimination of Wastes) and granted it the power to issue subsidies and loans, to conduct research on recycling, and to provide technical advice on the reuse of wastes.

The French government has not fully utilized the recycling provisions of the law of 1975, which depend upon administrative decrees and orders to become effective. The one major exception to the nation's reluctant implementation of the 1975 law is the waste oil program. A decree and orders issued in 1979[71] and modified in 1985[72] created an integrated management system that governs the handling of waste oil from ''cradle to grave.'' Its requirements are superimposed on the additional requirements of France's second major waste law, the law of 1976, which limits pollution at major industrial facilities, including rerefining plants.[73] The waste oil system was thus an early effort to implement the management authority of France's 1975 law, and it remains the one full-fledged example of the potential scope of that law as it might be applied to other kinds of recyclable materials.

The Obligations of Producers, Collectors, and Disposers

The decree and orders ensure the collection of all quantities of waste oil over 200 liters and their delivery to approved regeneration facilities. The definition of waste oil, and hence the scope of the waste oil program, is narrower than under German law. Waste oils are defined as ''used mineral or synthetic oils which are no longer suitable for the use for which they were intended as new oils, and which can be reused either as primary materials through recycling or regeneration, or as industrial fuel.''[74] This definition includes only those used oils that can be readily recycled as lubricants or fuel. Waste oils that fall within the German definition, such as metal-working oils, emulsions, various mixtures containing oil in low proportions, and oils that are discarded without being "used," are outside the scope of the French waste oil system and are governed instead by other waste disposal requirements.

Commercial generators of waste oils have distinct obligations under the French system. The 1979 decree requires every producer to turn his waste oil over to an approved collector or authorized disposer, unless he himself has obtained approval as a disposer.[75] In order to maintain the quality of the waste oil for purposes of recycling, the decree requires producers to keep waste oil separate from other wastes and to avoid mixing waste oil with other substances.[76]

A single collector is selected by authorities in each of the country's administrative districts *(départements)* to collect waste oil within that district.[77] Under the terms of the administration's approval, the collector must seek out potential producers of waste oil, publicize the availability of collection services,[78] and pick up all quantities of waste oil within the zone in quantities over 200 liters.[79] With the exception of certain industrial oils that can be reused with little or no processing, the collector is further obliged to deliver the collected waste oils to an approved facility.[80] Collectors must have adequate storage facilities that allow separation of different types and qualities of waste oil.[81] Collectors must also give receipts to producers for the waste oil collected[82] and must maintain records and report monthly to government authorities concerning the origins, amounts, and destinations of that waste oil.[83] For motor and industrial oils, the collectors must report their collection prices and, under new requirements added in 1985, the prices charged to recyclers for the delivery of waste oil.[84]

In return for meeting these obligations, the collector gains exclusive rights to collect all quantities of waste oil within his designated zone or zones.[85] Unlike the German system, which has relied only on subsidies to help finance the collection of small and otherwise unprofitable quantities of waste oil, the French system has granted collectors the benefit of a government-enforced monopoly, on the premise that the profits from the resale of larger quantities of waste oil in each zone will serve to subsidize other unprofitable but obligatory collection tasks.

Recyclers and disposers of waste oil are likewise required to receive approval from the Ministry of the Environment in addition to meeting pollution control requirements imposed under the environmental law of 1976. The decree and orders on waste oil management give priority to the regeneration of waste oils as lubricants and generally prohibit burning, with the narrow exception of waste oil that cannot be rerefined, which may be burned only in approved industrial facilities.[86] In practice the approvals issued for the management of waste oil have been limited almost exclusively to rerefining facilities. The license requires facilities to accept all waste oil offered to them within their capacity; to maintain records on dates, quantities, characteristics, and origins of waste oil received, and on rerefined lubricants shipped from the facility; to give collectors a written receipt stating the quantity and quality of each shipment received; and to meet all environmental laws and regulations.[87]

Allocation and Pricing of Waste Oil Supplies

Provisions governing the relationship between collectors and disposers underwent substantial modifications in 1985. The most important changes pertained to the pricing and allocation of the waste oils delivered to rerefiners and to the regulation of waste oil exports.

The 1979 collection system was intended to operate as a public utility, providing equal collection service to all producers and equal access to waste oil supplies by all rerefiners. Since the rerefiners' demand for waste oil exceeded availability, the 1979 decree empowered ANRED to allocate the collected waste oil based on the preexisting market shares of the rerefiners.[88] In addition to receiving their allocated shares the rerefiners remained free to compete for waste oil delivered directly to them by producers.

Under the 1979 system the collector for each zone was selected primarily on the basis of a bid, submitted as part of his application, proposing the price at which he would offer waste oil from his zone to rerefiners. The ministry granted approval to the collector in each zone who offered the lowest price for waste oil from that zone.[89] The collector was then bound, during the three-year term of the government's approval, to offer waste oil to rerefiners at a price no higher than the one he had proposed, adjusted as necessary for inflation.[90]

The 1979 system also provided for equalization of the collectors' charges. Since the amounts charged by individual collectors could vary according to their bids and according to the different costs of transportation from each zone to any given rerefiner, the goal of delivering waste oil at equal rates to all rerefiners in the country required a reallocation of expenditures among rerefiners. Under the 1979 system ANRED periodically calculated the amounts owed by rerefiners to each other to spread the costs of the collection system equally among them. The rerefiners in turn accomplished the redistribution of costs by making payments to a central fund administered by a trade organization.[91]

The government's control over the collection system and the allocation of waste oil supplies had the effect of precluding exports. The terms of the 1979 decree and orders allowed delivery of waste oil only to facilities approved under French law. The collectors consequently could not legally export the waste oil they obtained.

The price fixing and exclusion of exports proved to be controversial and difficult to enforce. The process of selecting collectors for their low bids guaranteed a low official price for waste oil delivered by the collection system, but it caused the collectors to offer only the minimum prices set by law to the producers of waste oil. These minimum prices were lower than the prices offered elsewhere in Europe for the same types and qualities of waste oil. In particular, they were substantially exceeded by the prices offered for waste oil as fuel in countries such as Belgium, where burning of waste oil in greenhouses and other small facilities was allowed and widely practiced.[92] The prices offered in Germany were also higher, a phenomenon attributed variously to the German waste oil subsidy and to economies of scale in the German rerefining industry.[93] Waste oil was therefore exported illegally, and it was also used and marketed semi-clandestinely in France for burning in small facilities.[94] Though contrary to law, burning waste oil was an attractive option, given the substantially higher prices of virgin fuel oil.

The regional agencies charged with enforcing the delivery of collected oil to rerefiners (the D.R.I.R., Directions Régionales de l'Industrie et de la Recherche) were unsuccessful in preventing the export and burning of waste oil. The D.R.I.R. had opposed the implementation of the 1975 law on waste disposal and resource recovery, and was frequently accused by the rerefiners of deliberately failing to enforce the exclusive rights of the authorized collectors and rerefiners.[95] The rerefining industry mounted an increasingly bitter attack on the administration for its perceived policy of overlooking infractions. Meanwhile, in the national courts, in the political arena, and in the press, organized groups of service station owners challenged the validity of the existing decree and orders that gave the authorized collectors and disposers exclusive rights to waste oil.

Proceedings were also initiated in the European Court of Justice to determine whether the monopolistic aspects of the French system violated the principles of free trade expressed by the Treaty of Rome in establishing the European Economic Community. In two decisions issued on February 7, 1985, the court concluded on the one hand that the French provisions prohibiting burning of waste oil comport with the provisions of the 1975 Waste Oil Directive and that a system based on authorization of collectors by zones does not violate principles of free trade established in the Treaty of Rome.[96] The court found, on the other hand, that the provisions of French law requiring delivery of waste oil to authorized collectors and authorized disposal facilities implicitly preclude exports, and it held that such restrictions on trade do violate the EEC Treaty.[97]

In response to all these difficulties, the French administration abandoned the price-fixing and anti-export provisions of the 1979 system in 1985. Legal requirements were formally amended to allow competition in the marketing of waste oil supplies, and the government's responsibilities for setting the price of waste oil supplies and allocating available quantities were formally abolished.[98] Under the 1985 amendments collectors and disposers are required to negotiate contracts among themselves. Waste oil may also be delivered to facilities in other member countries if those facilities have been properly authorized in accordance with the 1975 directive. The government selects and licenses collectors simply on the basis of technical qualifications, not on the basis of the prices offered.

While relinquishing its control over the pricing and allocation of waste oil supplies, the government has simultaneously taken steps to reassert its oversight of the waste oil management system. The monopoly rights of the authorized collector in each zone have been confirmed, and the administration has strongly restated its intention to take action against any infringement of those rights.[99]

To make up for the elimination of protectionist measures, the French administration also established a temporary new subsidy of the rerefiners. The waste oil industry had benefited from government support measures in the past, most recently a fund administered by ANRED and financed by a short-term tax on lubricants. The tax was collected from July 1, 1979, through December 31, 1981, and resulted in a fund of 73.28 million francs, which ANRED used in various ways to improve waste oil management through loans and investment subsidies to rerefiners, collectors, and producers, and through financing of public awareness programs and waste oil collection centers. Although the tax was collected only from 1979 to 1981, part of the fund remained unspent as of 1985.[100]

The administration initially resisted creation of a further subsidy, but faced with mounting pressure from the rerefiners and their supporters in the National Assembly, in 1985 it imposed a new tax on lubricants to benefit the rerefining industry. The goal was to compensate the French rerefiners for competitive disadvantages suffered because of the disparate implementation and enforcement of the 1975 directive in the member states of the European Communities. The subsidy was therefore approved as a temporary measure, to last only two and a half years, in the anticipation that other countries would soon take action to remove the competitive advantages of non-compliance with the directive.[101]

Assessing the Effectiveness of the French Waste Oil System

As a result of the reporting requirements of the decree and orders governing waste oil management, the French government has gained precise information on the amounts and destinations of all waste oil handled within the official system. This information provides a clear picture of the transportation of waste oil by authorized collectors and the regeneration of waste oil by rerefiners. The amounts delivered to rerefiners, predominantly black motor oils, have hovered around 100,000 tons per year, and small additional amounts have been burned in authorized industrial facilities.[102]

The widespread bypassing of the official system for purposes of burning waste oil means that the export and burning of waste oil is not well recorded. Estimates have typically put the quantities exported for burning at 30,000–40,000 tons per year and the quantities burned in France, likewise, at 30,000–40,000 tons per year,[103] but the president of the rerefiners' trade organization recently estimated that a total of 120,000 tons is exported or burned yearly.[104] The amount of waste oil that could be regenerated but is dumped illegally is generally put at around 30,000 tons per year.[105] Given the narrow definition of "waste oil" used in France, these statistics do not include some waste oils, particularly soluble oils and oily mixtures, that are covered by the EC Waste Oil Directive but fall outside the French waste oil system.

The available statistics, though incomplete, reflect a success in encouraging the recovery of waste oil. If one assumes that up to half of the motor oils used each year in France become waste oil then one may conclude that less than 15 percent is dumped, while more than 40 percent is delivered to the rerefiners for regeneration of lubricants, and the rest is burned in France or exported.[106] In the United States, by contrast, the statistics reflect a much higher level of dumping and disposal and an insignificant amount of rerefining.[107]

While the French statistics do not compare favorably with the German figures for waste oil delivered within the official management system, the French apparently have been much more successful in preventing the contamination of their waste oil supplies with other materials. Thus, while the Germans have struggled recently to develop a system to exclude contaminants from waste oil, the French government and industry widely assert that contamination is not a serious problem in France. As a result of the lower levels of contamination of waste oil, the average yield of regenerated lubricants from waste oil is said to be much higher in France than in Germany.[108] Although the potential for PCB contamination of waste oil has been a cause for concern, there has been no public furor over PCB levels in rerefined lubricants as in Germany. The major potential source of PCB contamination in France would be electrical transformers, most of which are in the hands of the national utility Électricité de France (EDF). The government is in the process of taking an inventory of such transformers and determining ways to ensure safe disposal.[109]

The narrowness of the French waste oil definition, and the resulting legal differentiation between recyclable materials and wastes, may account in part for the French success in preventing significant contamination of waste oil supplies. In addition, the tendency of the authorized collectors to specialize in waste oil and not to handle other hazardous materials may be influential.[110] The official system is primarily designed to collect and deliver waste oil supplies of adequate quality for rerefining, and the rerefiners thus serve as an effective screening point by analyzing the waste oil supplies they receive.[111] To the extent that waste oils enter the official recycling system in France, there is little incentive to mix in other combustible hazardous materials, given the prohibitions on burning.

LESSONS FOR THE UNITED STATES

The United States is far from achieving a collection and recycling program equivalent to that of France or Germany. As we have seen, many of the United States' waste oil resources are lost through indiscriminate dumping or careless handling, and many are burned in unregulated circumstances without air pollution monitoring or controls. The more highly developed European systems deserve special attention as the federal government considers measures to control waste oil management in the United States.

One striking aspect of the Waste Oil Directive, and the French and German systems implementing it, is the use of subsidies and incentives to encourage collection and prevent uncontrolled dumping. The U.S. government has resisted such market intervention. Yet the French and German experience shows that mechanisms for subsidizing or otherwise supporting the collection of waste oil can have distinct advantages over mere regulatory prohibitions against improper disposal. Effective enforcement of prohibitions is extremely difficult given the enormous number of oil users and the ease of improper disposal. Enforcement becomes especially problematic when waste oil prices are low and costs of collection or treatment far exceed the potential profits from recycling. Subsidies and incentives, by contrast, can serve to make collection an attractive option without the expense of vastly expanded enforcement efforts.

A system of subsidies or incentives can also have the advantage of enhancing the government's ability to oversee waste oil management. If collectors and disposers must keep records and submit reports in order to qualify for government support, the government readily gains information on the structure and functioning of the industry. The French and German governments have both gained detailed knowledge of waste oil handling by making benefits contingent on the submittal of information.

The main drawback of subsidies is that they risk supporting continued operation of inefficient enterprises that would otherwise be unable to compete. It is therefore important that a system of subsidies be sufficiently flexible or short-term to avoid substantial disruption of competition among firms. This is particularly true because changes in the markets for waste oil may quickly render a particular subsidy level out of date. Variable subsidies or short-term programs that take these changes into account make sense.

The German government's decision to gradually reduce subsidy payments expresses the view that a strong system, once established, may become sufficiently entrenched to survive without further government support. The French government's elimination of its earlier pricing system and the introduction of short-term subsidies also reflects the position that government intervention in the market may become less necessary or need alteration over time. The French and German experiences thus suggest that temporary subsidies and other short-term government support measures are useful means of helping the establishment of a waste oil management system, and that governments should pursue programs that take changing economic factors into account, rather than making long-term commitments to finance the recycling industry.

Another prominent feature of the Waste Oil Directive and the French and German systems is the licensing of collectors and disposers and the record keeping on waste oil shipments. While these requirements resemble ''cradle-to-grave'' waste management, they are not part of the hazardous waste disposal programs, and indeed the success of the waste oil programs depends on maintaining separate systems for waste oil and hazardous waste. In the United States, a controversy over whether waste oil should be regulated as a hazardous waste has mired federal efforts to license collectors and track waste oil shipments. Yet, as France and Germany have recognized, strict licensing and reporting requirements are needed to preserve waste oil as a resource free of contamination and to ensure its recycling. Both countries have successfully required authorization of collectors and disposers and record keeping on shipments as an integral part of sound waste oil management.

Finally, the French and the Germans, in implementing the Waste Oil Directive, have addressed the potential for air pollution from burning waste oil. Both countries limit combustion to licensed industrial facilities where air pollution can be monitored and controlled. Waste oil prices in France have been low because of the virtual prohibitions on burning, but the less sweeping restrictions in Germany have not had the same effect. The German experience suggests that waste oil burning may be limited to industrial settings and controlled through permit conditions without seriously undermining the value of waste oil as fuel. The Commission of the European Communities, too, in proposing amendments to the Waste Oil Directive, recognized the advisability of limiting waste oil burning to large facilities where air emissions can be controlled. Given the wide variety of possible contaminants in waste oil, such restrictions are wise precautions that deserve consideration in the United States.

NOTES

1. The United States Environmental Protection Agency determined not to regulate used oil destined for recycling under the Resource Conservation and Recovery Act of 1976, but it reserved judgment on whether used oil destined for disposal might be regulated in this manner. Regulations on the sale of used oil for burning were promulgated on November 29, 1985.

2. 75/439/EEC, adopted by the Council of the European Communities on June 16, 1975, amended by 87/101/EEC, adopted on December 22, 1986.

3. An excellent discussion of these various regulatory considerations can be found in W. A. Irwin, "Used Oil: Comparative Legislative Controls of Collection, Recycling, and Disposal," *Ecology Law Quarterly* 6 (1978): 699.

4. Environmental Resources Limited, "Implementation of Directive 75/439/EEC on the Disposal of Waste Oils," Final Report, prepared for the Directorate General for Environment, Consumer Protection, and Nuclear Safety, Commission of the European Communities, September 1983, p. iii.

5. United States Environmental Protection Agency, "Composition and Management of Used Oil Generated in the United States," EPA/530–SW–013, November, 1984, 1–4, 1–5.

6. Ibid., 2–3 and 3–10, on sources and types of waste oils.

7. Ibid., 1–10 to 1–17, 3–6 to 3–8, and 3–26 to 3–40, on the additives and contaminants in waste oils.

8. Ibid., 1–5 to 1–10, 4–1 to 4–25, and 5–1 to 5–24, on the various ways that waste oils are handled and disposed of.

9. The directive as issued in 1975 defines "waste oils" very broadly. Under former article 1, the term means "any semi-liquid or liquid used product totally or partially consisting of mineral or synthetic oil, including the oily residue from tanks, oil-water mixtures and emulsions." Under the 1986 amendments, the term is more narrowly defined to mean "any mineral-based lubrication or industrial oils which have become unfit for

the use for which they were originally intended, and in particular used combustion engine oils and gearbox oils, and also mineral lubricating oils, oils for turbines and hydraulic oils.''

10. Articles 2 and 4.

11. Article 3.

12. Article 4.

13. Article 6 and article 13, as amended (former article 12). The Waste Oil Directive uses the term ''disposal'' more broadly than the Environmental Protection Agency's regulations under the Resource Conservation and Recovery Act, which draws distinctions among ''storage,'' ''treatment,'' and ''disposal.'' Former article 3 of the Waste Oil Directive confirmed that ''disposal'' included ''recycling (regeneration and/or combustion other than for destruction).'' Under article 1, as amended, the term means ''the processing or destruction of waste oils as well as their storage and tipping above or under ground.'' The directive also refers to ''holders'' and ''collectors''—the counterparts of ''generators'' and ''transporters'' in American statutes.

14. Articles 11 and 12, as amended (former articles 10 and 11).

15. Article 5 as amended provides: ''Where the objectives defined in Articles 2, 3 and 4 cannot otherwise be achieved, Member States shall take the necessary measures to ensure that one or more undertakings carry out the collection and/or disposal of waste oils offered to them by holders, where appropriate in the area assigned to them by the competent authorities.'' Former article 5 contained similar language.

16. Article 14 (former article 13).

17. Article 15 (former article 14).

18. Articles 14 and 15 (former articles 13 and 14).

19. The fund established by the Comprehensive Environmental Response Compensation and Liability Act of 1980, 42 U.S.C., sections 9601 et seq.

20. Benno W. K. Risch, ''Die Abfallwirtschaftspolitik der europäischen Gemeinschaft, insbesondere auf dem Altölgebiet,'' in *Proceedings of the First European Congress on Waste Oils,* September 1976, published by the Commission of the European Communities, Directorate-General ''Scientific and Technical Information and Information Management,'' pp. 287–307.

21. Article 1, as amended. See footnote 9 above.

22. Articles 7, 8, and 10, as amended. ''Toxic and dangerous waste'' is defined by Directive 78/319/EEC.

23. Articles 7 and 8.

24. Article 8, as amended. The emissions limits apply to plants with a thermal input of 3 MW or more, while plants with less thermal input must be ''subject to adequate control.''

25. Article 3, as amended: ''Where technical, economic and organizational constraints so allow, Member States shall take the measures necessary to give priority to the processing of waste oils by regeneration.''

26. Gesetz über Massnahmen zur Sicherung der Altölbeseitigung [Law concerning Measures to Assure the Disposal of Waste Oil, hereinafter Waste Oil Law].

27. The Waste Oil Law was amended by laws of May 4, 1976, and October 24, 1979. It was superseded by Gesetz über die Vermeidung und Entsorgung von Abfällen (Abfallgesetz or Waste Law) of August 27, 1986.

28. Bundesamt für gewerbliche Wirtschaft.

29. Waste Oil Law, section 3(1).

30. Waste Oil Law, section 3; Bekanntmachung über die Festsetzung der Pflichtgebiete und die hinterlegten Preislisten nach dem Altölgesetz vom 7. Marz 1984. Erste Verordnung zur Durchfuhrung des Altölgesetzes, section 4.

31. Waste Oil Law, section 3(2).

32. Waste Oil Law, sections 1 and 2.

33. Statistics on the costs of recycling enterprises were last gathered in 1980, but the yearly reductions of 1 DM in the subsidy did not begin until 1984. "Fünfter Bericht der Bundesregierung über die Tätigkeit des Rückstellungsfonds nach dem Altölgesetz, insbesondere über die Möglichkeiten einer Ermässigung der laufenden Zuschusse und der Ausgleichsabgabe," Deutscher Bundestag, 10. Wahlperiode, Drucksache 10/1229, April 3, 1984, pp. 6–8.

34. Ibid., appendices 6 and 7.

35. United States Environmental Protection Agency, "Composition and Management of Used Oil Generated in the United States," EPA/530–SW–013, November 1984, pp. 1–8 to 1–10.

36. For example, the German statistics cover all "waste oil" under the German Waste Oil Law, while the Environmental Protection Agency's report is limited to "used oil."

37. Gesetzentwurf der Bundesregierung, Entwurf eines Vierten Gesetzes zur "Änderung des Abfallbeseitigungsgesetzes," Deutscher Bundestag, 10. Wahlperiode, Drucksache 10/2885, February 2, 1985, p. 18.

38. Gesetz über die Beseitigung von Abfallen (Abfallbeseitigungsgesetz), section 1(3).

39. There are four such groups or committees, formed to address coordination of enforcement of environmental laws in the eleven states or *Länder* and in Berlin. The four committees deal, respectively, with issues of air, water, waste, and noise.

40. R. Berghoff, chairman of the LAGA subcommittee, interview.

41. The Conference of Environmental Ministers (Umweltministerkonferenz [UMK]). The UMK makes recommendations on behalf of the *Länder* for new federal legislation.

42. R. Berghoff, interview.

43. Ibid.

44. H. Schnurer at the Ministry of the Interior gave an estimate of 75 percent; R. Berghoff, chairman of the LAGA subcommittee, gave an estimate of 66 percent.

45. C. Lafrenz of Dr. Dr. Anton Maier AG, interview.

46. R. Berghoff, interview; C. Lafrenz, interview.

47. G. Alfke of the trade group MWV (Mineralölwirtschaftsverband e.v.), interview; memoranda of MWV dated April 4, 1985.

48. G. Alfke, interview.

49. Waste Law § 5a(1). Waste oil is subject to the law even if it would not otherwise fall within the more general definition of waste in the statute. The earlier definition of waste oil is retained in the new law. Waste Law § 5a(1).

50. Waste Law § 3.

51. Waste Law § 11, ¶ 2.

52. Waste Law § 14.

53. P. Kromarek, Institute for European Environmental Policy, Personal communication, May 27, 1987.

54. Waste Law § 12.

55. Waste Law § 30(4).

56. Waste Law § 5b.

57. Waste Law § 30(2).

58. R. Berghoff, interview.

59. R. Berghoff, interview; C. Lafrenz, interview.

60. A. Szelinski of the Umweltbundesamt, interview. C. Lafrenz at Maier expressed the view, however, that current handling of old transformers by scrap metal dealers remains a source of ongoing contamination as well.

61. H. Schnurer, interview.

62. P. Plechatsch, Federal Office for Trade and Industry, interview; R. Eder, Bundesverband Privater Sonderabfallbeseitiger und Rückstandsverwerter e.v. (BPS), interview.

63. Loi du 13 septembre 1940, as amended by Loi n° 365 du 15 juillet 1943 (J.O., September 19, 1940, and July 18, 1943), Obligation de récupérer et de régénérer les huiles minérales de graissage; Loi du 1er octobre 1940 (J.O., October 15, 1940); Loi du 9 novembre 1940 (J.O., November 14, 1940).

64. Arrêté du 20 novembre 1956, as amended by Arrêté du 11 mai 1957 (J.O., November 22, 1956, and May 21, 1957).

65. Circulaire du 14 mai 1976 des ministères de la qualité de la vie et de l'industrie, Interdiction de brûlage des huiles usagées.

66. Article 23 de la loi n° 80–531 du 15 juillet 1980 relative aux économies d'énergie et à l'utilisation de la chaleur: "Les seules utilisations des huiles minérales et synthétiques qui, après usage, ne sont plus aptes à être utilisées en l'état pour l'emploi auquel elles étaient destinées comme huiles neuves, . . . sont, lorsque la qualité de ces huiles usagées le permet, la régénération et l'utilisation industrielle comme combustible. Cette dernière utilisation ne peut être autorisée que dans des établissements agréés et lorsque les besoins des industries de régénération ont été préférentiellement satisfaits" (unofficial translation by the author).

67. Loi n° 75–633 du 15 juillet 1975 relative à l'élimination des déchets et à la récupération des materiaux (hereinafter Law of 1975).

68. Law of 1975, article 8.

69. Law of 1975, articles 9 and 20.

70. Law of 1975, articles 10 and 21.

71. Décret n° 79–981 du 21 novembre 1979 portant réglementation de la récupération des huiles usagées (hereinafter "1979 decree"); Arrêté du 21 novembre 1979 relatif aux conditions de ramassage des huiles usagées (hereinafter "1979 order on collection"); arrêté du 21 novembre 1979 relatif aux conditions d'élimination des huiles usagées (hereinafter "1979 order on disposal").

72. Décret n° 85–387 du 29 mars 1985 modifiant le décret n° 79–981 du 21 novembre 1979 portant réglementation de la récupération des huiles usagées (hereinafter "1985 decree"); Arrêté du 29 mars 1985 relatif aux conditions de ramassage des huiles usagées (hereinafter "1985 order on collection"); Arrêté du 29 mars 1985 modifiant l'arrêté du 21 novembre 1979 relatif aux conditions d'élimination des huiles usagées (hereinafter "1985 order on disposal").

73. Loi n° 76–663 du 19 juillet 1976 relative aux installations classés pour la protection de l'environnement.

74. "Les huiles usagées concernées par le présent décret sont les huiles minérales ou synthétiques qui, inaptes après usage à l'emploi auquel elles étaient destinées comme huiles neuves, peuvent être réutilisées soit comme matière première en vue de recyclage

ou de régénération, soit comme combustible industriel. . . . '' 1979 decree, article 1 (unofficial translation by the author).

75. 1979 decree, articles 2 and 3, amended by 1985 decree, article 2.

76. 1979 decree, article 2.

77. 1979 decree, article 4; 1979 order on collection, articles 2, 3, 4, as amended by 1985 order on collection, articles 3, 4, 5.

78. 1979 order on collection, appendix article 9; 1985 order on collection, appendix article 7.

79. 1979 order on collection, appendix article 14; 1985 order on collection, appendix article 8.

80. 1979 order on collection, appendix article 14.

81. 1979 order on collection, appendix article 10; 1985 order on collection, appendix article 13.

82. 1979 order on collection, appendix article 12; 1985 order on collection, appendix article 10.

83. 1979 order on collection, appendix article 14; 1985 order on collection, appendix article 8.

84. 1985 order on collection, appendix article 8. The collector must now simply publish the prices at which he will buy waste oil, and he must report those prices to the administration. 1985 order on collection, appendix article 7.

85. 1979 order on collection, appendix article 17; 1985 order on collection, appendix article 15.

86. 1979 decree, article 7, amended by 1985 decree, article 6.

87. A burner of used oil, if not classified as a major industrial facility falling within the terms of the law of 1976, must nevertheless meet pollution control requirements. 1979 order on disposal, appendix article 2; Décret n° 74–415 du 13 mai 1974 relatif au contrôle des émissions polluantes dans l'atmosphère et à certaines utilisations de l'énergie thermique; Arrêté du 21 mai 1980, Equipements et exploitation des installations thermiques consommant des huiles usagées (J.O., June 7, 1980).

88. 1979 order on disposal, article 2. The quotas were arrived at by dividing the currently available resources by the market shares of each enterprise as they existed two years before the regulatory system was established. C. Duday of ANRED, interview.

89. 1979 order on collection, appendix article 7. Although the collector's ability to guarantee compliance with environmental requirements was a relevant criterion under the law, in practice the price offered for waste oil was the determinative factor in selecting collectors under the 1979 system.

90. The formula for adjusting prices for inflation was set forth in the 1979 order on collection, appendix article 15.

91. SAHU (Société d'Approvisionnement en Huiles Usagées).

92. Le Matin, August 27, 1985, reported a difference of 400 F per ton between the market price in France (600 F) and in neighboring countries (1,000 F). France exports more waste oil than any other member country in the European Communities, and it has no imports, other than very small quantities from Luxembourg. See Commission of the European Communities, ''Proposal for a Council Directive Amending Directive 75/439/EEC on the disposal of waste oils,'' COM(84) 757 final, January 24, 1985.

93. Cidecos Formation, ''Rapport sur les possibilités de maintien d'activités sur le site de Gargenville,'' January 1984, p. 29; C. Duday, interview.

94. See ANRED, Comité de gestion de la taxe parafiscale sur certains lubrifiants mis à la consommation, Rapport d'activité 1983, p. 5.

95. P. Brassart, President of the Chambre Syndicale du Reraffinage, interview.

96. Case 240/83, February 7, 1985.

97. Case 173/83, *Commission of the European Communities v. French Republic,* no. C 59/03, February 7, 1985.

98. 1985 decree, article 5; 1985 order on disposal, article 1.

99. Circulaire du 3 avril 1985 du Ministère de l'environnement, Direction de la prévention des pollutions.

100. C. Duday, interview.

101. F. Combrouze, Ministère de l'environnement, interview.

102. ANRED, Comité de gestion de la taxe parafiscale sur certains lubrifiants mis à la consommation, Rapport d'activité 1983.

103. Environmental Resources Limited, "Implementation of Directive 75/439/EEC on the Disposal of Waste Oils," Final Report, prepared for the Directorate General for Environment, Consumer Protection and Nuclear Safety, Commission of the European Communities, September 1983, p. 58; Cidecos Formation, "Rapport sur les possibilités" pp. 29–30.

104. M. Perrimond, *Le Matin,* August 27, 1985.

105. Cidecos Formation, "Rapport sur les possibilités," pp. 29–30.

106. ANRED publications assume that 65 percent of motor oils in France are consumed during use and only 35 percent arise as waste oil. ANRED, "La Récupération et le recyclage des huiles usagées," 1984, p. 6. A recent industry study, however, suggests that this figure is out of date and that arisings of 45–50 percent are more plausible (Cidecos Formation, "Rapport sur les possibilités," p. 29.) Based on the 1983 motor oil use of 428,000 tons, estimates of arisings for that year thus range from 149,000 to 214,000 tons.

107. See U.S. Environmental Protection Agency, "Composition and Management of Used Oil Generated in the United States," EPA/530–SW–013, November 1984, pp. 1–8. More than 40 percent of the waste oil held by "do-it-yourself" motorists and "automotive generators" is dumped or discarded.

108. P. Brassart, interview.

109. M. Richard of the D.R.I.R for l'Ile-de-France, interview.

110. The collection industry includes two large companies with 80 percent of the market that are directly affiliated with rerefiners. SRRHU (Société de ramassage pour la régénération de huiles usées) is affiliated with the rerefiner SOPALUNA, while COHU is affiliated with the rerefiner MATTHYS. On the structure of the French waste oil market, see Cidecos Formation, "Rapport sur les possibilités," pp. 30–33. The small family-run operations that hold the remainder of the market may be changing among the independent collectors, however. J. Martin, waste oil collector, interview.

111. J. Dumortier, MATTHYS-Lubrifiants, interview.

7

Waste Havens and Waste Transfers: International Transboundary Issues _____

BRUCE PIASECKI AND WENDY GRIEDER

In Europe, a shipment of hazardous waste crosses a national frontier every five minutes—on a 24–hour clock. The significance of this frequency cannot be neglected. As a result, most European nations now firmly acknowledge the need for international cooperation in regulating waste transfers. The issues that receive primary international attention include the siting of facilities near national borders, the role of cooperation in the joint design of new treatment centers, and the disposal or incineration of hazardous wastes at sea. But an equally important international issue is the transboundary movement of high-risk wastes; for within this deceptively simple issue—mixing transport, storage, and transfer risks—the fate and efficacy of most treatment markets reside. Americans have much to learn on securing such markets across political boundaries.

Both European and North American countries have extensive experience in the transboundary movement of hazardous waste. Hundreds of thousands of legitimate shipments take place safely every year across national frontiers, but the possibility always exists that the waste will become lost or will not be properly managed along its rather involved route. The United States, for example, has had its share of export plans that involved problems. A proposed export of PCBs to Honduras for incineration hinged on the use of a fraudulent document to convince EPA to make an exception to the U.S. Closed Border Policy for PCBs. Another shipment of hazardous waste to Costa Rica was unexpectedly returned to the United States by Costa Rican authorities because of the country's inability to adequately treat the waste. The exporting company subsequently sued EPA, and one of its employees, for prejudicing the shipment. A judge denied a request for injunctive relief filed by the company to take care of the waste sitting on a dock in Long Beach, California; later, the entire suit was dropped. These examples illustrate that the movement of hazardous waste is truly a global issue. Each year the number of international shipments increases dramatically, not only within Europe but throughout North America as well.

Why are hazardous wastes sent from their place of generation to another nation? Sometimes there is simply a lack of facilities in the country of origin that can adequately treat the waste. Some generators in the United States, for example, send their hazardous waste to Europe because of special European approaches for reclaiming materials that derive considerable commercial value from the wastes. Another major reason for exporting hazardous waste is public pressure to treat the waste elsewhere. Two years ago a Chicago company went bankrupt and was left with a large amount of highly toxic cyanide chips. The company had closed because the managers were under indictment for murder in the death of an employee who had handled the chips. The potential disposal of this material in the Chicago area created intense controversy, and there was enormous pressure to treat the chips elsewhere. The company proposed to go to Canada; but when the nature of the waste and the news of the employee's death reached Canadian officials and the public, the issue got so hot they ultimately refused to accept the waste across their border. It was finally treated at a facility in Cook County on order by the court.

The third major reason for exporting hazardous waste, examined in Chapter 5, is an economic one. As the impact of stronger regulation is more severely felt, whether it is in the United States or in Europe, the costs rise proportionately, thus encouraging a waste generator to look for cheaper prices elsewhere. By the time a generator absorbs the costs of storage, transport, treatment, and the liabilities to cover these activities, countries with less regulation and resulting lower costs look increasingly attractive to hazardous waste generators. At times these three reasons are strong enough to transfer considerable amounts of complex, high-risk wastes, and only uniform international agreements among the involved nations are sufficient to properly monitor these complicated activities.

THE MOST ACTIVE INTERNATIONAL ORGANIZATIONS

The Organization for Economic Cooperation and Development (OECD) is the most active international organization in this area. The OECD grew out of the Marshall Plan, first designed to assist in the economic recovery of Europe. It currently has a membership of 24 industrialized countries. The present mission of the OECD is to promote economic growth of member countries, to help less developed countries, and to encourage trade expansion across the globe. While the environment was not specifically included in the OECD Convention signed in 1960, ten years later the OECD established an Environment Committee to place more emphasis on the qualitative aspects of economic growth. This includes intelligently addressing the problems of air and water pollution, as well as the problematic set of issues surrounding the management of chemicals and hazardous waste.

In 1976 OECD adopted a recommendation that called for member countries to have comprehensive hazardous waste management policies. At that time there was a growing awareness that hazardous waste was being generated in increasing amounts in member countries. Moreover, OECD also acknowledged that economic and social pressures were resulting in the frequent export of waste to member and non-member countries. This led to the conclusion that it was necessary for the OECD to establish a coordinating system that matched national obligations for the monitoring and control whenever wastes crossed national frontiers. Readers are encouraged to view these developments as matured responses to the intristic characteristics of waste management noted in our introduction.

The United Nations Environment Program (UNEP) was formed in 1972 as an outgrowth of the U.N. Conference on the Human Environment (the Stockholm Conference). Its broad mandate is to protect the human environment by seeking solutions to pollution and man-made contamination, and to promote environmentally sound economic and social development in both urban and rural areas. UNEP is governed by a 58–member Governing Council, but there are over 80 nations that participate in various UNEP activities, including many topical hazardous waste management issues and controversies.

UNEP has long recognized that waste management differs substantially in different regions of the world, particularly according to their state of economic development. This significant imbalance demands a heightened cooperation between nation-states, especially regarding the safe transfer of hazardous waste. Thus the UNEP hazardous waste activities are designed to provide frameworks for effective and environmentally sound hazardous waste management policies in radically different cultural and economic settings. This is especially important in areas under rapid development, such as the Far East and Asian nations, where protecting the environment is not a top priority. In addition to the transfrontier issue, the UNEP mandate provides for a wide range of information-transfer activities among nations.

The Commission of European Communities (EC), also known as the Common Market, has a membership of twelve European nations. This network has joined together to carry out a four-point mandate: to live peacefully together, a goal better achieved through international cooperation; to be able to negotiate with the superpowers from a position of united strength; to enable member country economies to develop in conditions of fair competition; and to improve living conditions and ensure social progress.

It is now over a decade since EC heads of state formally acknowledged that the economic growth inspired and fostered by the Community had to be linked with improvements in living standards, the quality of life, and protection of the environment and natural resources. The result has been the search for a common European environmental policy that aims to balance economic revitalization with environmental protection. Many of the waste reduction strategies described in chapter 2 have thereby received positive press and support within EC networks.

EC's shared environmental policy often focuses on the problems of hazardous waste. The Community recognizes that more and more hazardous wastes are crossing national borders, significantly increasing the potential for inadequate management or deliberate mismanagement. As a result, the Community Council has adopted a series of directives controlling the transboundary movement of hazardous waste within the European Community. Although not a member of the Community, the United States works closely with the EC on this issue, both through regular meetings and bilateral contacts with the Community's staff.

Under the North Atlantic Treaty Organization Committee on the Challenges of a Modern Society (CCMS), the United States, Canada, and several European member countries are cooperating on a major information exchange on dioxin and related compounds. The three-year pilot study, jointly led by the United States, the Federal Republic of Germany, and Italy, consists of three related work groups. One focuses on risk assessment, another on technology assessment, and the third on the accidental release of dioxin into the environment. There is increasing interest in dioxin both as scientists discover more about its properties and effects and as more dioxin sources are identified. One of the major benefits of this information exchange is the elimination of duplication in this resource-intensive area of research. Participating countries will identify the ongoing efforts in Europe and North America and then initiate new research initiatives where there are shared needs. Countries also now have shared access to monitoring data collected under special conditions, such as the long-term plant and tissue studies in the Seveso area and the treatment technologies for dioxin-contaminated soil being used at Times Beach, Missouri.

RECENT TRANSFRONTIER AGREEMENTS: CLOSING THE GAPS

The OECD Waste Management Policy Group's (WMPG) major project involves transfrontier movements. To date it has been the subject of two binding OECD Council decisions and a series of influential recommendations. These decisions mandate that member countries engage a coherent waste management system in a manner available to member scrutiny, and that they implement a binding agreement on the standards for safe transboundary movements by the end of 1987. This agreement is envisioned to include a prenotification of exports, use of an OECD manifest to track the waste, and some harmonization of export regulations and procedures in member countries.

The latter is easier said than done. One of the thorniest hurdles for international agreements remains the definition of hazardous waste, which still varies widely from country to country. This has led to extensive discussions of what wastes should be subject to the OECD convention. As a result, OECD countries have agreed that if a waste is considered hazardous in any country involved in any part of the transaction (exporting, importing, or transit country), it will be covered by the procedures. This important move has spawned a complex system of cross-referencing national lists of hazardous waste. (The results of these harmonization efforts have been summarized by Bruce Piasecki and OECD consultant Harvey Yakowitz and are available by writing the American Hazard Control Group.)

Liability during a transfrontier movement is also a pressing issue. The OECD has begun to examine in detail the role of insurance, compensation, and financial guarantees in the context of transboundary movement of hazardous waste. However, it has drawn no binding conclusions at this point nor has a decision been made on whether this work should proceed. There is no firm disagreement among members that the liability issue should be addressed, yet most representatives believe it is so complex an issue that attention to it will slow work on the rest of OECD's projects concerning improvements in waste management.

The OECD is also deeply concerned about the export of hazardous waste to non-member countries. OECD countries have recognized that the potential for environmental, health, and foreign policy risks is significantly greater for exports to non-member countries. Therefore, they have made this issue a central part of their conventions. At its March meeting in 1986 the WMPG approved a draft OECD Council decision calling for non-member countries to issue prior written consent to a shipment before the exporting country can allow the shipment to take place. It also requires exporting countries to halt a waste shipment if it is not satisfied that the waste will be adequately taken care of, even if the receiving country consents. These important developments exemplify the central role of ethical standards in transboundary issues. Without such moral sanctions, prevention of inappropriate shipments proves impossible.

An exciting recent development entails guidelines issued by the Environmental Law Program of the United Nations Environment Program. These guidelines, designed for use by developing countries, cover waste management from generation to disposal with special emphasis on the transboundary movement of hazardous waste. A set of "Cairo Guidelines" has been completed recently under this initiative that will be offered for adoption at the UNEP Governing Council meeting in the spring of 1987.

The next step in this UNEP project is to hold a series of technical meetings aimed at assisting developing countries in establishing their own national hazardous waste management systems. This ambitious implementation project includes building an infrastructure to administer a waste management plan, as well as assisting nations in evolving reliable regulations and policies for the transport, storage, and disposal of hazardous waste. Lessons from the European experience in building an effective treatment infrastructure, as examined in chapter 5, should prove instructive for this UNEP effort.

The Commission of European Communities has also been active in the management of hazardous waste, particularly concerning international shipments between member countries. The result has been the development and implementation of an EC directive for transboundary shipments, which is binding on all member nations. This directive calls for prenotification using a standard EC

form, use of an EC manifest, as well as packaging, labeling, and liability requirements. Although the United States is not a member of the EC, it is subject
to the provisions of the directive should U.S. generators export their hazardous
waste to EC countries.

PERENNIAL PROBLEMS IN WASTE TRANSFER CONTROLS

Strategies that monitor exports and imports of high-risk wastes are not without
their serious shortcomings. One problem remains the efficient harmonization of
the elements of the different national systems so that a country and its regulated
community are not unnecessarily burdened by complying with separate procedures for wastes crossing national frontiers. Efforts to achieve compatible manifests for the OECD and the EC, for example require the same kinds of reporting
information that suits the needs of member countries of each organization. The
OECD and EC are also working to develop regular shared notification procedures
including identifying the same competent authorities in a given country to administer the system and to respond to crises or accident.

Individual governments experience problems when they lack the authority to
meet these international requirements. This often necessitates changing legislation, which, depending on the structure of the government, can be a complicated
and taxing process. If the United States, for example, fails to comply with the
binding decisions of the OECD under the new American export regulations, the
process to legislate and promulgate rules could take well over a year and perhaps
two. Although there is a rich concensus in Europe and North America for many
of the actions of the OECD and EC, select countries are now faced with making
extensive changes in their export policies, which may drastically slow the implementation of desired multilateral systems for the transboundary movement of
wastes.

Despite all of these international efforts, there still are considerable problems
with transfrontier shipments in Europe, as well as in the United States. There
are substantial incentives for business to operate outside the system; countries
thereby will continue to suffer instances of hazardous waste dumping, particularly
in nations with minimal controls. Therefore, rigorous nation-state enforcement
must remain the linchpin for each industrialized nation. However, if effectively
managed and kept simple, multilateral systems for tracking waste for international
shipments can also go a long way in reducing the risks associated with the
management of hazardous waste by simply preempting confusion and articulating
shared standards.

THE BENEFITS OF INCREASED U.S. PARTICIPATION IN
INTERNATIONAL WASTE CONTROL AGREEMENTS

Through its participation with OECD, the United States has gained useful
knowledge on European regulatory approaches to enforcement, the siting of

facilities, and the development of new treatment technologies. The latter is becoming increasingly important as the United States implements more regulations requiring alternatives to land disposal. Moreover, the United States has much to learn from the European countries in collection of waste from small-quantity generators, something that is regularly practiced in several European nations and assessed in detail in Chapter 10. Many European countries make it easier for generators to comply with the regulations by facilitating the collection of waste and ensuring a regular access to treatment and handling facilities. By contrast, many American generators suddenly find themselves without a reliable source of transport or disposal because of the rapid changes in the law or the fluctuating costs of transport, disposal, or insurance.

The United States itself has played a major role in dealing with international shipments of hazardous waste. The flow of innovation is often transatlantic in both directions. The United States, for example, was at the forefront of the government-to-government prior notification system for exports. Moreover, as a result of a recent amendment to the Resource Conservation and Recovery Act (RCRA), U.S. regulations for hazardous waste export are significantly stronger. Europe has benefited from this U.S. experience; several European countries now have hazardous waste export policies similar to that of the United States.

The new RCRA regulations, which took effect in November 1986, require an expanded notification from the exporter and a receiving country's prior written consent before any shipments can take place. This gives the United States an enforcement mechanism should the consent not accompany the shipment, and it will give countries, particularly developing ones, the opportunity to say no to an export. This is especially important if a country determines that it lacks the capability to adequately manage the incoming wastes.

These new regulations provide a positive signal to the international community. They suggest that the United States takes seriously its responsibility to ensure that its exports of waste are properly handled and that the receiving countries are adequately prepared. The European nations are watching with interest the implementation of these new regulations, since it is possible that these standards may be adopted in Europe, either bilaterally or regionally, with only minor changes.

There is also a new provision in the U.S. regulations for negotiating bilateral agreements with countries. Meant to eliminate some of the more burdensome administrative sections of the legislation, this provision allows for the legitimate trade in hazardous waste where exports are mutually beneficial to the two countries. For example, the United States and Canada have a great deal of trade in hazardous waste across a common border. Under the general framework of the proposed agreement, the United States would notify Canada about all intended shipments. Canada would then inform the United States if the shipments would be allowed to take place, and vice versa. Under the OECD convention on the transboundary movment of hazardous waste, it is envisioned that European countries will make similar bilateral or regional agreements in which they agree on procedures for prior notice, manifesting, and transport.

These changes would, of course, improve the overall marketability and transferability of wastes currently undervalued in certain regions either due to oversupply or lack of sufficient treatment capacity in a given region. This increased marketability, the major benefit of opening wastes to international competition for treatment, should also improve the management and ultimate destiny of some of industry's worst wastes. For without access to diverse treatment facilities, some treatable wastes are mistakenly forced into the ground where they only constitute the most suspect form of waste havens. Yet as noted in Chapter 5 closing borders along national boundaries may also prove a politically needed gesture in some circumstances to sustain a nation's treatment infrastructure.

Part 3

The Challenge Ahead

8

Siting Hazardous Waste Facilities: Asking the Right Questions

GARY DAVIS AND WILLIAM COLGLAZIER

The siting of hazardous waste management facilities presents modern democracies with one of their most difficult tasks. Vehement public opposition leading to the failure of many contemporary siting attempts is common lore in the United States. Nonetheless, the siting problem remains a critical obstacle to improved hazardous waste management in America, since without adequate treatment capacity the nation will not move beyond land disposal. Moreover, safe reuse and treatment processes will remain lost in American laboratories.

There have been many prescient analyses of the siting predicament.[1] Many American states now have elaborate siting procedures that incorporate extensive public participation, preemption, siting criteria, enforced bargaining, compensation schemes, or some combination of these (Appendix 4 reports on some of these state siting initiatives). Most of these procedures, however, ignore a set of fundamental questions implicit in the European approach that must be addressed before the American public will become satisfied that a particular facility should be located in a particular location.

These fundamental questions concern basic strategies for hazardous waste management, such as the demonstration of the need for new facilities, the social choice of technology, the communication of the risks and uncertainties of these technologies, and the effectiveness of conventional regulatory systems in policing the proposed facilities. Since a consensus about these fundamental questions is rarely attempted before siting processes begin, siting proposals inevitably get bogged down in these concerns. Once persistently misperceived as irrelevant by regulatory agencies and facility proponents, this set of questions now demands sustained responses.

With the exception of Sweden, the European nations studied offer little direct insight on the issue of what process is best to use to site unwanted hazardous waste facilities, but they may have a lot to teach us about answering the fundamental questions that preceed siting. At least until recently, the combination of government ownership with the use of incineration and treatment technologies in integrated facilities has been a successful answer to these questions for several European nations. There is evidence, however, that some of the consensus about these fundamental issues has broken down in Europe, and that this traditional European approach may no longer be acceptable to the public. Nonetheless, the comprehensive planning process by which European nations sited superior facilities should prove instructive to Americans, especially as we experience the rapid closure of land disposal.

HAZARDOUS WASTE FACILITY SITING IN EUROPE

Most of the European hazardous waste facilities discussed elsewhere in this book did not face significant public resistance when they were sited, partly because they were sited before the current public awareness about the potential dangers of hazardous waste had arisen. The disasters at Love Canal and the Dutch Lekkerkerk dumpsite were unknown to the world when the Germans sited the Schwabach facility in 1968 and the Ebenhausen facility in 1972, or when the Danish sited the Kommunekemi facility in 1975.

Furthermore, the basic premise behind public ownership—namely, that safe hazardous waste management is a public necessity—appears to have at least resolved some of the preliminary concerns that always disrupt the American siting process. The perception that the government-owned facilities were being designed with the best available technologies and with the minimization of risk foremost above profit may have alleviated some of the public concern about the siting of these earlier facilities (indeed, the idea that government-owned facilities would be easier to site was one that prompted the chemical industry in Hessen, West Germany, to join in the formation of HIM, the quasi-public hazardous waste management company in Hessen).

More recent siting efforts in Europe have encountered difficulties, however, and most European nations are in the same siting quandary today as we are in America. In an effort to enhance the current debate, this section will review some of the more recent siting experiences in Europe. By emphasizing the siting of the integrated treatment and disposal facility in Norrtorp, Sweden, this section examines a siting process that goes well beyond most of the processes that have been initiated in the United States both in its attention to health and safety concerns and in its deliberate pursuit of meaningful public input.

The Swedish Public Involvement Plan

As described more fully in Chapter 5, SAKAB is the Swedish government-owned hazardous waste management company. Formed in 1976, the company's single purpose is the construction and operation of hazardous waste management facilities. Planning began in 1976 for a large integrated treatment facility, and a site was chosen in 1978 by the management of SAKAB. In February 1978 an application for a license to construct the facility was submitted to the National Franchise Board for Environmental Protection, which issues all licenses for projects that may have a significant impact upon the environment.[2]

The site chosen for the facility was near Norrtorp in the central part of southern Sweden. It was in a predominantly industrialized and fairly remote area that had already been subject to extensive groundwater contamination from previous mining operations. The nearest community is approximately one mile away.[3]

The National Franchise Board is structured like a court of law with a deliberately mixed set of members. The president has legal training, and others on the board include a technical expert, a representative from an environmental organization, and one member from either industry or local government, depending on the type of facility that is being considered. Like the U.S. National Environmental Policy Act, the Swedish Environment Act requires that a license application for ''environmentally dangerous activities'' must include an assessment of all of the environmental impacts of the activity. This also includes siting considerations. The board must notify the public about the application, consult with the affected local governments, provide an opportunity for public comment, and may hire its own experts to make special investigations at the expense of the applicant.[4]

The public quickly became actively opposed to the siting of the SAKAB facility and held several meetings and demonstrations. Members of the opposition included local government officials, environmental organizations, and members of the general public. Despite this opposition the Franchise Board granted the permit for the facility in December 1978. This decision did not end the siting dispute, however, since the decision of the Franchise Board was appealed by the National Environmental Protection Board, the central regulatory agency for environmental matters, and by local intervenors. The National Environmental Protection Board felt that the Norrtorp site did not have convenient enough access to a receiving stream of sufficient size to accommodate its treated water discharges.[5]

To resolve this intragovernmental dispute, the minister of agriculture set up a committee for the purpose of identifying and describing alternative sites for the location of the SAKAB facility, focusing mostly on coastal options. The committee evaluated seven sites during 1979, but when these were compared with the initial choice of Norrtorp, they were each considered to be less suitable for the facility. In the summer of 1980 SAKAB proposed yet another coastal site as an alternative, but the National Franchise Board rejected this application since it considered the Norrtorp location to be more suitable.

Finally, in January 1981 the national government, which has the last word on siting and licensing decisions, decided that the facility should be built in Norrtorp and that active public involvement should be further sought. In addition, the government strengthened the licensing requirements to take into account some of the concerns about health and safety. The public involvement plan consisted first of the appointment of a "Consultative Group" of community representatives to have input into the design and operation of the facility.[6] Members of the group, who were appointed by the County Council, include representatives from the following organizations:

1. The county branch of the Farmer's Association
2. The Fisherman's Association for Lake Hjalmaren, into which the aqueous effluent of the plant ultimately flows
3. The Sports Fisherman's Association of the county
4. The Water Conservancy for Lake Hjalmaren
5. The County Property Owners' Association
6. The County Tenants' Association
7. A local environmental organization
8. The Working Group against SAKAB (the group numbered about 100 people)
9. The county branch of the Swedish Trade Union Confederation
10. The county branch of the Swedish Central Organization of Salaried Employees
11. The Swedish Confederation of Professional Associations
12. The Swedish Employers' Confederation

One of the first tasks of the Consultative Group was to draw up a preoperation environmental survey plan for the area surrounding the facility. This was to be performed before the facility began operating. The survey looked at air quality, soil conditions, surface water quality, groundwater quality, concentrations of metals in vegetation, and concentrations of organic contamination in cattle, fish, and wildlife. The group also met several times to study some of the design and operating features of the facility and was able to call in its own experts. Some of the specific concerns discussed by the group included groundwater conditions in the area, the effect of water discharges on fishing, and the potential public health effects of PCB incineration. The group continues to function in an oversight capacity for the facility.[7]

In addition to the survey, the more stringent license that resulted from the siting dispute included a requirement for a trial operation period. The trial period began with the start-up and operation of the facility in 1983 and continued for over a year until the end of 1984. To evaluate the efficiency of the facility, the Franchise Board then appointed an independent consultant who resided at the facility and extensively monitored it under a full range of operating conditions. The consultant was also given the power to shut the facility down or to prohibit the incineration of certain types of hazardous wastes if he felt that this posed a significant risk to human health. In addition, the Research Institute of the Swedish National Defense Agency also performed a risk analysis focusing on reducing the risks of fires, explosions, and spills, resulting in several changes in the design of the plant.

At the end of the trial operating period the consultant prepared a report for the Franchise Board, and following a careful review of this report, the board issued its final operating permit for the facility in October 1985, seven years after the original application. In the final license, the Franchise Board also required SAKAB to secure a 500–meter buffer zone around the facility by purchasing property within this distance. Only three pieces of private property were within this distance, and one person living 200 meters away decided to remain on his property.[8]

The community is still not happy with hosting Sweden's major hazardous waste facility, but the opposition has decreased now that the trial period is over and the final permit has been issued. The Consultative Group has given many of those opposed to the facility an opportunity to have their health and safety concerns addressed in the design and operation of the facility, which may have lessened the resistance. The lengthy siting process was costly for SAKAB, but SAKAB officials believe that the consultative process and the trial operation period resulted in a safer facility.[9] American policymakers should take note of the broad range of representation in the Consultative Group, as well as its powers to review design and performance standards. The notion of a trial operating period where many parties evaluate performance with the threat of a shutdown in mind is equally compelling, given the intensity of recent conflicts over siting of incineration facilities.

Siting Hazardous Waste Facilities in West Germany

In West Germany siting decisions about industrial facilities, including hazardous waste facilities, are a state government planning function conducted by the regional office of the state environmental and planning agency. Siting decisions include public hearings and a public comment period and are appealable to a court of law.[10]

There have been two hazardous waste facilities recently constructed in West Germany, both of which encountered public opposition. These are the Biebesheim incineration facility owned by HIM, the quasi-public hazardous waste management company for the state of Hessen, and the ZVSMM Raindorf landfill, used for disposal of treatment residues from the Schwabach integrated treatment facility. A third facility, a conventional landfill proposed for Mainflingen in the state of Hessen by HIM, has been postponed indefinitely for the testing of alternative technologies for long-term storage of treatment residues. (See Chapter 5 for more information about HIM and ZVSMM.)

The HIM incineration facility near Biebesheim was first proposed in about 1979 and is located in an industrial area with a refinery on one side and a chemical plant on the other. The plant met active opposition, including demonstrations at the site, from residents in the surrounding area before being approved in 1980. Opposition to the facility has been minimal since the start-up of the plant in 1982.[11]

The landfill proposed by HIM for Mainflingen has been the subject of extreme controversy for over four years. Planned for an abandoned clay pit, the landfill was designed to accept treatment residues and a limited amount of untreated hazardous waste from industries within Hessen. Currently, Hessen is without landfill capacity within the state for the treatment residues from the HIM facilities, and most of these residues are shipped to other West German states for disposal. Concerns about the impact of the landfill on drinking water sources in the area prompted vehement opposition to the facility and helped the Green Party, the political wing of the environmental movement in West Germany, secure seats in the state parliament.[12]

Because of the shift in politics in the state, the construction of the landfill was postponed. The ruling Social Democratic Party in the state has now formed a coalition with the Green Party, and part of the agreements leading to the coalition dealt with the Mainflingen landfill. Leaders of the two political parties agreed that the Mainflingen landfill will not be built as originally designed, and that HIM will instead demonstrate an above-ground design that will be more protective of groundwater resources. Furthermore, some of the residues that were to be deposited in the landfill will instead be sent for retrievable storage deep underground in salt mines near Herfa Neurode, also in Hessen (see Chapter 3 for more information about this repository).[13]

The Raindorf landfill recently sited by ZVSMM, the publicly owned hazardous waste management company for the district of Middle Franconia in the state of Bavaria, resulted from a painstaking site suitability study for the district. The siting may have been aided by a small measure of "compensation" provided by ZVSMM to the host community. In the two-year-long site suitability study, the geology, groundwater resources, population distribution, and transportation routes of the district were surveyed to narrow the search down to twenty sites for more thorough evaluation. After further evaluation of each site, the Raindorf site was chosen as being the most suitable because of its geology, the size of the buffer zone, and the easy access to major highways. There was some local opposition to the site, resulting in an appeal of the licensing decision to a court of law. The decision of the planning authorities to license the facility was upheld by an administrative court, and no further appeal was filed.[14] As a private measure of compensation to the local community where the landfill was sited, ZVSMM used its equipment to construct a sports arena for the community.[15]

Siting of Finland's Integrated Treatment Facility

One of the newest integrated treatment facilities in the world is the Oy Suomen Ongelmajate (Finnish Problem Waste Management Company) facility constructed in 1982–1984 near the city of Riihimaki in southern Finland (see Chapter 5 for more information about the Finnish Problem Waste Management Company). The law setting up the quasi-public company set out some general siting requirements for the facility: the plant should be in southern Finland, highway access should be good, rail access should be available, the geology should be suitable for a landfill capable of receiving treatment residues, and the site should be reasonably near a municipality with a district heating system so that the energy production of the incineration system could be utilized. The search for the site of the plant lasted over a year until Riihimaki was selected as meeting the required criteria.

There was strong resistance to the siting of the facility at the location where it was originally proposed, which was about two miles from the center of Riihimaki, a town of 24,000 inhabitants. The resistance led to the selection of a new location approximately four miles from the center of town.[16]

An Analysis of the European Situation

Clearly, siting high risk facilities differs vastly in different political cultures. The previous consensus on the management scheme for hazardous waste has been eroded by the disasters of the late 1970s and 1980s. Even if concerns about the hazards of incineration and land disposal of residues could be mollified with changes in designs and added safety features, the equity issue, which relates to the size of facilities and their geographic scope of operation, would remain a paralyzing source of public opposition in Europe.

The examples cited above also show that government ownership of facilities does not automatically improve public acceptance of hazardous waste facilities anymore, if it ever did. Although one of the principal reasons that waste generators in the West German state of Hessen agreed to participate in the establishment of HIM as a quasi-public hazardous waste management company was the belief that government-owned facilities would incur less public resistance in siting, industrial officials now concede that this belief has not proven true in Hessen.[17] After going through the process of siting the Raindorf landfill, officials of ZVSMM now express the opinion that government ownership in and of itself does not significantly aid in the siting of facilities.[18]

Nonetheless, a solid advantage of government ownership may be a greater willingness and ability to be flexible in addressing the health and safety concerns of the public. The Swedish example shows that, once a siting decision is made, extensive consultation with the public on health and safety concerns can both improve public acceptance of the facility and improve the facility's design and operation. The Finnish example also showed similar flexibility in the government's agreement to move the site two miles from the location originally selected. Few private companies would be able to spend the additional capital required to meet such public concerns in the United States.

The controversy over the Mainflingen landfill in Hessen, moreover, shows the importance of establishing a consensus on the technology to be used before attempting to site a facility. During a lengthy planning and site selection process the consensus about a particular technology selected may erode. Again, in the Mainflingen example the quasi-public company was able to be flexible in changing its technical approach to meet public concerns.

In none of these siting efforts did compensation prove a significant factor in reducing public opposition. Although several European communities near publicly owned incineration facilities receive inexpensive heat from incinerator heat recovery systems for their district heating systems, this was not identified as a significant factor in public acceptance. More overt efforts at compensation, such as the building of the sports complex by the ZVSMM management, may have won over some key local officials, but explicit compensation did not play a major role in any cases studied.

Finally, the equity issue was a sticking point in each of the siting efforts discussed. People simply objected to having their community serve as the host for the treatment or disposal of hazardous waste from all over the country or region. This fact alone counters some of the arguments in favor of large centralized facilities. Economies of scale and the regulatory benefit of having a central point of control are irrelevant if the public will not accept the centralization of risk in a single community.

ROADBLOCKS TO SITING: THE ROOTS OF THE PROBLEM

Before we take up the issue of how to design an equitable and effective siting process, there are some basic lessons that can be derived from observance of siting attempts in Europe and the United States. These may help identify the roots of the siting problem. Any siting process must appropriately respond to these concerns or it will likely fail.

1. *Equity:* It is inherently unfair to foist the impacts of managing the waste of large numbers of people and industries upon a small group near the site of a facility. No amount of reassurances about safety will fully resolve this inequity and the sustained public resistance that it generates.

2. *Preemption:* Preemption of local authority for siting of hazardous waste facilities is a myth, especially in the United States, where there are many opportunities built into the system to stop a facility. The community will utilize the political process, the courts, and if necessary, resort to civil disobedience to stop an unwanted facility.

3. *Compensation:* Where people are strongly concerned about health impacts of waste management, economic benefits to their community will rarely prove sufficient to overcome public opposition.

4. *Experts:* An uninformed, inexpert public quickly becomes expert when faced with the prospect of a hazardous waste facility. Members of the public may end up knowing more about the potential risks of a given proposal than the regulators or the proponents of the facility.

5. *Basic Policy Issues:* To the chosen community, a siting proposal reopens all the basic policy and planning questions, such as the choice of technology for

the facility, how acceptable risk is defined, or the manner in which the site was selected. Since members of the community were generally not involved in the initial resolution of these issues, these concerns are just as important in the siting process as whether the facility meets certain minimum regulatory standards. For the decision makers to refuse discussion of these issues only causes further antagonism.

6. *Enforcement:* Because of the questionable track record of regulatory agencies in enforcing hazardous waste regulations and performance standards at existing hazardous waste facilities, the public is understandably very concerned about enforcement of safety standards for a hazardous waste facility being sited in their community. This concern cannot be ignored in the siting process.

7. *Back Door Approaches:* Facility siting cannot be achieved any longer by rigging the approval process or by minimizing public information and public participation. The public reacts negatively to even the best facilities if a back-door approach is tried.

ASKING THE RIGHT QUESTIONS: TURNING ROADBLOCKS INTO ROADMAPS

The design of a hazardous waste facility siting process that takes basic policy issues into account can avoid the siting gridlock that currently dominates most American efforts. In these failed attempts, the issue of siting of hazardous waste facilities is usually framed too narrowly, thereby parking the process from the start. Decision makers too often ask, "How can we get people to accept an unwanted hazardous waste facility in their community?" By framing the issue in this fashion, we are actually asking the last question first.

As illustrated by the road blocks listed above, there are more fundamental questions that must be answered before attempting to site an unwanted hazardous waste facility in a particular community. The community must be convinced that these questions have been dealt with satisfactorily before the actual siting process can begin. These questions include the following:

Has the Need for a Facility Been Demonstrated?

The demonstration of need is particularly important. This question goes to the very core of any hazardous waste policy, since it is not self-evident that generators of hazardous waste are entitled to send hazardous waste somewhere away from its point of generation to be managed. Moreover, it is not self-evident that hazardous waste must be generated in the first place, as seen by the successes in waste reduction that are becoming more widespread throughout industry.

Demonstrating the need for a facility is extremely difficult because of a dearth of information about hazardous waste generation, existing management methods, and future trends in waste reduction. In the United States the need issue is usually ignored, since it is presumed that a firm will not invest the capital in a hazardous waste facility unless the market exists. It is also presumed that the consequences of a failure to accurately predict that market are of no concern to the public, but merely affect the bottom line of the firm. This ignores the fact that financially pressed facility operators often cut corners and may use illegal or unsafe practices to try to remain profitable. On the other hand, if market signals are uncertain, a shortage of safe hazardous waste management capacity may lead to a continuation of inadequate land disposal practices. In the United States this potential treatment shortfall has been worsened by the increase in the quantity of regulated wastes under the 1984 RCRA amendments and by the closure of most of the interim status landfills in the country.

The centralized, government-owned management systems in Europe may not be any more adept at predicting need for facilities, but the consequences of an overestimation of the market are mitigated both by government ownership and the non-profit status of the facilities. The consequences of underestimation can be remedied in an orderly fashion by constructing new facilities. Furthermore, if the need issue is dealt with in a public forum rather than a corporate boardroom, the potential host community may feel more secure that the facility is truly needed.

The idea that need should be demonstrated before certain types of facilities are permitted to be built is not totally foreign to the free-market system in the United States. Public utilities, such as gas and electric companies, must generally prove "public necessity" before being permitted to invest in new facilities. Some states in the United States, such as New Jersey, have performed needs assessments and pursued organized plans for securing their "needed" hazardous waste facilities (see Appendix 4 for a discussion of siting processes in the United States). Although these approaches generally do not prohibit the siting of facilities that are not needed, they approach the need issue directly in a public way.

What Is the Appropriate Size and Geographic Scope of Operation of a Facility?

The appropriate size and geographical scope of operation of a hazardous waste facility is related to the equities of siting and may prove one of the central siting issues. Although there may be economies of scale in building large, centralized hazardous waste treatment facilities and other benefits acquired from centralizing management functions and enforcement, bringing hazardous waste into a single community from a wide region almost guarantees opposition. Smaller, more specialized facilities located in the industrial communities they serve incur less resistance in siting because they are inherently more equitable.

If communities are given the responsibility at the local level for planning for the management of hazardous waste generated by their industries and the resources and information to do so, new facilities that may be needed would be sited with the equity issue of scale in mind. It may be that local governments would join together to site a regional facility as they did in the District of Middle

Franconia in Bavaria, West Germany. The fact that this decision to centralize would come from the ground up rather than from the top down should ease the resistance based upon equity.[19]

A further step in the direction of community responsibility is to ensure that facilities would first serve the needs of the community by allowing restrictions on import of wastes from other areas. Although state and local laws prohibiting the import of wastes from other areas have been struck down as violating the commerce clause of the U.S. Constitution, allowing greater local control over imports of hazardous wastes may actually result in the siting of more hazardous waste facilities, since the facilities would be perceived as more equitable.[20]

The answer of many of the larger hazardous waste generators in the United States to a perceived shortage of qualified hazardous waste management facilities has been to treat more hazardous wastes on-site. The uniquely American response of the hazardous waste management industry to siting difficulties and the demand for on-site treatment has been the development of transportable treatment units that bring the treatment process to the waste rather than the waste to the treatment process. Transportable units are only in a community long enough to deal with the wastes generated by a particular generator in that community and thus reduce perceptions of inequity generated by permanent treatment facilities accepting waste from wide areas.

Is the Technology Selected Appropriate?

Technology selection brings into the debate questions about risk and scientific uncertainty. Does the technology selected pose unacceptable or unknown risks? Who decides whether or not these risks are acceptable?

The siting process is not technologically neutral. Although different technologies present different risks, too many facility proponents and regulatory agencies assume that the public is not sophisticated enough to understand the technical issues at stake or the questions of acceptable risk that are implicitly tied to the choice of technology. Attempting to sidetrack these issues with blithe assurances of safety, without truly addressing and comparing health and safety aspects of alternatives, inevitably backfires.

It is unlikely that a community where a hazardous waste facility is proposed had any real participation in rule-making decisions where such issues were decided. Moreover, where most of the decisions about choice of technology have been left to the marketplace, public involvement is sidetracked completely. At the point of a siting proposal, an important choice faces the facility proponent and the regulatory agency. They can either attempt to foreclose these issues by pointing to standards already set in stone, or they can openly consider and actively encourage the input of the community on these critical issues.

As was the case for SAKAB in Sweden, the genuinely open and flexible approach to these technical issues can both restore public trust and result in construction of safer facilities. Along these lines, the ability of the community to hire independent experts to evaluate the claims of safety made by facility proponents would be an important improvement in the siting process. A fund should be available to potential host communities for these independent evaluations.

Another important mechanism for overcoming the scientific uncertanties of technology selection is to establish a trial operating period accompanied by extensive monitoring during actual operating conditions. As noted above, this approach was practiced at the SAKAB facility in Sweden quite successfully. Such an evaluation must be truly independent, with the assurance that if the evaluation of the facility shows that the risks are more significant than anticipated, the facility will be shut down until fundamental design changes are completed or permanently if no such changes would reduce risk to an acceptable level.

Can the Public Be Sure That the Facility Will Be Operated Safely and in Compliance with Regulations?

This is a question that no regulatory agency wants to deal with, since to admit that it is a valid question is to admit that the hazardous waste regulatory system relies extensively on voluntary compliance. It is virtually impossible for overburdened hazardous waste regulatory agencies to adequately police the thousands of hazardous waste generators, the thousands of hazardous waste shipments, and the hundreds of treatment, storage, and disposal facilities, both on-site and off-site. The public has firm grounds to assume that violations will occur and that the regulatory agency will probably not detect or correct many of these violations.[21] The daily news diet of environmental disasters at old disposal sites and the slow response of government agencies also fuels this wariness.

Since this underlying lack of faith in our regulatory system exists, it must be dealt with before the facility siting process can proceed. One strategy is to provide mechanisms for community involvement in the oversight of hazardous waste facilities. The Consultative Group for the SAKAB facility in Sweden is a good example of how this might work. The group included the most vehement opponents of the facility. Another approach is to allow local government officials to exercise supervision over the facility in conjunction with state regulators.

Does the Decision-Making Process Sufficiently Allow the Views of Key Stakeholders to Be Heard and Taken into Account?

It is only after the first four questions are dealt with that the actual decision-making process for hazardous waste siting becomes relevant. Regardless of who makes the final siting decision, the decision process itself must be perceived by the public as not just a sham exercise to comply with minimum formal rules on public participation. The four preliminary issues discussed above ideally should be addressed in a public forum prior to designation of particular sites. In order for the answers to the questions above to carry any weight, designated communities must perceive that their interests were represented in the resolution of these issues.

The development of a management plan by the agency involved in the siting decision is an effective means of addressing the preliminary issues discussed above. A small policy dialogue group representing the key interests and stakeholders can help to point to where consensus may lie on certain controversial issues before a broader public participation effort is undertaken. The general public can then be given an opportunity to comment on the management plan after a good deal of outreach and educational efforts to acquaint the public with the key issues.

Whether the siting process involves a preemptive decision by a state regulatory agency or siting board or some sort of bargaining and negotiation between the facility proponent and the potential host community, the purposes of a broad public participation effort are twofold: (1) to receive citizen input in order to improve the decision and (2) to increase the likelihood of public acceptance of the decision by having a process that is recognized as open and fair. To achieve this dual purpose, public information about the siting process should be communicated early and widely, using different media, understandable language, and in manageable amounts. It should also come from as many credible sources as possible. Key people should be involved at critical points in the process, not only formally through hearings and negotiations, but also informally through meetings, workshops, and dialogue groups. The process should not promise a great deal and then fail to deliver: it should be realistic about funding, staffing, workloads, and timeliness, and about how open it will be. Ironically, it may not matter who ultimately makes the final siting decision as long as the process is viewed as a fair one.

Is the Actual Site Selection Process Sufficiently Objective and Scientifically Credible?

One problem with the laissez faire approach to the siting of hazardous waste facilities is that the actual site selection is seldom made in an objective and scientifically credible manner. If the public knows or suspects that a site has been selected mainly for political reasons or because the land just happened to be available and inexpensive, opposition can be guaranteed. The same can be said for a selection process that attempts to rationalize a political decision after the fact by marshalling the scientific justification.

To be objective, a site selection process must make use of scientifically based siting criteria that take into account all of the potential impacts of each type of hazardous waste facility. The current federal hazardous waste regulations in the United States, which most states have copied, generally presume that if design, construction, and operation standards are adhered to, a hazardous waste facility can be sited virtually anywhere. This is a major weakness in America's siting protocol. Many states in the United States have gone a step further and created site selection processes using objective criteria. These criteria usually recognize the different hazards associated with different types of hazardous waste management technologies (Appendix 4 discusses some of these state developments in detail). EPA was charged by the 1984 RCRA amendments to develop federal siting criteria and is in the early stages of the regulatory process.

Can Incentives Be Provided That Would Adequately Compensate a Host Community for Negative Impacts That Are Not Associated with Health and Safety?

It is only after safety issues have been resolved that the issue of compensation for the host community can be brought up. Compensation that is offered too early in the process is viewed by the public as a distasteful attempt at bribery.

There are some legitimate uses for compensation. There may be a decline in property values in the area near the facility. Community services, such as fire protection, may need to be improved in order to accommodate the facility. Streets may need to be widened to allow for more truck traffic. Other negative impacts, such as noise, dust, and a general decline in the reputation of the community and its attractiveness for economic development may deserve compensation. The bargaining over compensation should be done on an equal footing, which probably means that either the community or the developer should have recourse to some higher authority if they fail to agree.

CONCLUSION

John Rawls, in his *Theory of Justice,* comments that ''justice is the first virtue of social institutions,'' but notes that different political systems can have different concepts of justice.[22] In the United States, our system is built on the conception of justice as fairness. Americans do not like to think that they are taking on—or giving to someone else—more than a fair share, especially when what is being shared is risk, not benefit. But the siting of hazardous waste management facilities leads ineluctably to an unequal distribution of risk: many people benefit from one process or another that generates hazardous waste, but only a few communities end up with a hazardous waste facility as a neighbor.[23]

Uneasy as Americans are with the notion of a process that has unequal results, they are even more uneasy with what may appear to be a thinly veiled utilitarian concept of justice: the greatest good for the greatest number. Although public decisions entailing the sacrifice of individual interests to broader social benefits are not unprecedented (such as our tax system or land condemnation for public projects), these decisions, although inequitable, generally produce recognized costs, not risks of arguable proportion; generally involve money or real property, not potential health impacts; and usually affect only the current population, not future generations as well. For hazardous waste facilities, none of these three characteristics holds, so vehement opposition based upon the equity issue is understandable.

The equity issue, then, remains the strongest, most formidable link despite the fact that different parties and perceptions constantly pull at that link. Technical experts supporting a proposed facility say, "The public is ignorant"; waste generators say, "The public is ungrateful"; regulatory officials say, "The public is unreasonable"; while the public says, "Why us?" Although the technical experts may be confident that the facility is needed and the real risks are indeed small, the potential host community must be convinced that this is the case and that it is not being unfairly imposed upon.

The most difficult challenge is to mobilize to a more widely held sense, both in the potential host community and in our broader social and political community, of the need for hazardous waste facilities. Superior equipment exists. The knowledge to operate it is available, as is the desire on the part of industry to use it to extinguish their liabilities. Yet if the discussed consensus is never attempted or achieved, continued opposition to hazardous waste facility siting can be expected however elaborate our formal siting procedures become.

NOTES

1. See Michael O'Hare, Lawrence Bacow, and Debra Sanderson, *Facility Siting* and *Public Opposition* (New York: Van Nostrand Reinhold, 1983); David Morell and Christopher Magorian, *Siting Hazardous Waste Facilites: Local Opposition and the Myth of Preemption*, (Cambridge, Mass.: Ballinger Publishing Company, 1982).

2. "This is SAKAB," information packet from SAKAB, Norrtorp, Sweden, 1985.

3. Lars Llung, Plant Manager, SAKAB, Norrtorp, Sweden, personal communication, July 23, 1985.

4. Swedish Environment Protection Act, 1969, sections 11–17.

5. Olov von Heidenstam, Director, Technical Department, National Swedish Environmental Protection Board, written communication, June 11, 1985.

6. "This is SAKAB."

7. Ibid.

8. Ibid.

9. Lars Llung, Plant Manager, SAKAB, Norrtorp, Sweden, personal communication, July 23, 1985.

10. Pascale Kromarck, Institute for European Environmental Policy, personal communication, Bonn, Federal Republic of Germany, June 1985.

11. G. Erbach, Plant Manager, HIM, Biebesheim, personal communication, August 1983.

12. Hans Christoph Boppel, member of Hessen Parliament, Green party, personal communication, July 11, 1985.

13. Excerpt from the "Red/Green Waste Agreements" in *Konzepete und Methoden der Sonderabfall Beseitigung*, Protokoll des Wiesbadener Sonderabfall Seminars, Die Grünen im Hessichen Landtag.[*Concepts and Methods for Special Waste Management*, Proceedings of the Wiesbaden Special Waste Seminars, The Greens in the Hessian State Parliament.] (June 23, 1984), pp. 3–5.

14. Hans-Georg Ruckel, Director, ZVSMM, personal communication, Schwabach, Federal Republic of Germany, July 8, 1985.

15. Franz Defregger, Director, Waste Management, Bavarian Ministry for Land Planning and Environmental Protection, personal communication, July 9, 1985.

16. F. Aarnio, "The Finnish System," paper presented at the 1st International Symposium on Operating European Centralized Hazardous (Chemical) Waste Management Facilities, Odense, Denmark, September 1982.

17. Helmut Schaum, Hoechst AG, West German Chemical Industry Association, personal communication, Frankfurt, Federal Republic of Germany, July 10, 1985.

18. Hans-Georg Ruckel, Director, ZVSMM, personal communication, Schwabach, Federal Republic of Germany, July 8, 1985.

19. See the discussion of the California and Florida approaches in Appendix 4.

20. Jonathan T. Cain, "Routes and Roadblocks: State Controls on Hazardous Waste Imports," *Natural Resources Journal* 23 (Ocotber 1983): 768–93.

21. *Illegal Disposal of Hazardous Waste: Difficult to Detect or Deter* (Washington, D. C.; U.S. General Accounting Office, Resources Community and Economic Development Division, 1985); Andrew Szasz, "Corporations, Organized Crime, and Hazardous Waste: An Examination of the Making of a Criminogenic Regulatory Structure," *Criminology* 24 (February 1986): 1–29.

22. John Rawls, *A Theory of Justice* (Cambridge, Mass.: Belknap Press of Harvard University Press, 1971), p.3.

23. For a parallel discussion of the equity issue in radioactive waste facility siting, see E. William Colglazier and Mary R. English, "Low-Level Radioactive Waste: Can New Disposal Sites be Found?" *Tennessee Law Review* 53, No. 3 (Spring 1986).

9

Government's Aid: The Role of Citizen and Environmental Groups in Europe

BRUCE PIASECKI AND JANET BROOKS

Citizen groups play an increasingly vital role in pursuing the proper enforcement of America's hazardous waste laws. National non-profit environmental organizations such as the Environmental Defense Fund and the Natural Resources Defense Council, combine the skills of lawyers and scientists to oversee the efforts of state and federal agencies. A key concern of these groups, and of a wide range of parallel state and local groups such as the Connecticut Fund for the Environment and the Gulf Coast Coalition for Public Health, is the implementation of toxic waste statutes already passed by Congress.

A classic role of American citizen groups is to force a reluctant or delayed EPA to act. When Congress mandates that EPA publish standards, this is usually only a small first step in an elaborate process. Getting EPA to fulfill congressional directives is often an uphill battle. For a variety of reasons—budget restrictions, organizational inertia, a habit of deferring difficult decisions, or, at times, basic disagreement with congressional will—the EPA has stalled frequently on toxic waste rule making since Love Canal. (Readers interested in a full-scale assessment of the strengths and weaknesses of the conventional American approach to environmental rulemaking should see: Piasecki's *Beyond Dumping*, 1984).

Although Congress's own oversight authority can be quite forceful, as seen by the rounds of resignations at EPA during President Reagan's first term, congressional scrutiny of particular regulatory or administrative decisions is often fairly limited. Congress is simply too busy to scrutinize fully the details of an agency's efforts. In recognition that state and federal agencies alone could not perform the monumental tasks of environmental protection, Congress gave citizens and citizen groups the formal authority to take legal action to force EPA fulfillment of Congress's stated requirements. More and more frequently this authority is used in "deadlines litigation"—lawsuits filed on behalf of citizen groups by professional environmentalists that force EPA to promulgate regulations by a certain date.

In contrast, there is a more frequent reliance on informal and non-judicial means for resolving environmental disputes in Europe. Whereas American citizen groups often directly challenge the substance of an EPA rule by requesting a full disclosure of the documents in question, European governments generally afford their citizens less formalized access to government documents. Moreover, while American citizen groups have the authority to act as private attorney generals or as surrogate enforcement officials by directly suing those who violate EPA regulations, group access to the courts is often less available in Europe. In Germany, for instance, only affected residents and neighborhoods, not groups organized around a cause, have standing in the court.

In response to these differences, European environmental activists and concerned citizens have evolved a rich set of alternatives to litigation. As America moves into an era of environmentalism that seeks less adversarial encounters on such divisive issues as acid rain and transboundary pollutants, this tool box of European alternatives assumes a special importance. What follows describes and assesses the salient features of these alternatives in four countries: West Germany, Holland, Great Britain, and Italy.

Throughout, a rough approximate distinction is made between citizen groups and environmental groups. Often local in scope and usually single-issue oriented, citizen groups lack the centralization of expertise evident in environmental groups—which are here defined as broad-based professional organizations that often tackle issues not related to the personal or local concerns of their participants. This distinction is inexact, although across the course of this essay it may serve to indicate why Europe has sprouted far more citizen groups and far less of America's card-carrying environmental groups.

ENVIRONMENTAL ACTIVISM AS POLITICS: THE WEST GERMAN EXAMPLE

A large number of local citizen groups are active in Germany, responding to problems associated with the inadequate or improper management of toxics. Lead contamination,[1] potential emissions from municipal incineration,[2] and a municipal landfill leaching dioxin (2,3,7,8–TCDD) into ground and surface waters[3] are leading items on the agenda. Since there is no functional Freedom of Information Act in the Federal Republic of Germany, German activists have developed considerable savvy with the press, using it to secure documents otherwise unattainable.

The Chemical Manufacturers Association in Germany believes that there will be no further problems caused by the current disposal of toxic wastes.[4] (The association does admit, however, that there will be continuing problems from older, abandoned dumps.) It bases this optimistic belief on increased regional awareness of the 1972 federal waste law. Since this law gives authority to the states to regulate their own region without setting uniform methods nationwide, one state can choose to ban the land disposal of a certain chemical while another state might allow it to continue.[5]

A representative from Germany's national environmental agency admitted that toxic waste controls would be strengthened if there were a more uniform approach. Yet he believes that uniformity will be brought about in Germany over time without the need for a formal legal document mandating national standardization. This bold confidence in the efficacy of peer pressure is noteworthy. He based his confidence on the small size of the country and its closely knit regulatory community, claiming that what one state does eventually causes a ripple effect throughout the country. A kind of regional peer pressure exists in Germany that motivates other states, in an effort to protect their own reputation, to meet standards upgraded elsewhere in the FRG.

The reputation of the state weighs heavily on German environmental decision makers. This informal weight is most palpably evident in West Germany's new management-specific waste catalog (see Appendix 3 for the excerpts on management of hazardous wastes). As the industrial world's first manual that concisely identifies the kinds of waste management that should be used for each of the nation's dominant waste streams, this influential guidebook is the product of competitive regional pride.

Starting in mid-1984, experts representing each of the states met to report on their region's most successful means for managing common industrial wastes. Each representative also reported on the management techniques his region felt patently inappropriate for certain wastes. Then, in June of 1985, on a day-long meeting that isolated all the key players on a cruise on the Rhine, the experts compiled their results. According to Carl Otto Zubiller, the facilitator of the meetings, "Peer pressure was at a peak. After fifteen years of separate regional approaches and investments, there was much to compare. Since the people of Hessen had decided not to place any halogenated organics in their soils, the obvious question was, What did they do with it? Once answered, the next question was, Why can't we do that?" Eventually, by a process of balanced mutual scrutiny, the overall plan was upgraded and extended.

Designed as a quality guidance manual, the catalog is binding on the state governments. "It's the weight of precedent that counts here," continued Zubiller. "No manager from any region wants to be known by the other *Länder* for using prehistoric or primitive management techniques."[6] West Germany's catalog, based neither on technology-specific rule making nor on top-heavy legal mechanisms, articulates shared management goals for an astonishingly diverse and vital industrial nation (see Appendix 3 for more details on West Germany's waste catalog).

This force of reputation also works internationally. Whereas competition among the states might be thought of as an invisible magnet that informally concentrates common developments, FRG's waste managers also scrutinize peers in their larger international context. When 41 barrels of dioxin-contaminated waste from the Seveso explosion disappeared en route to France in 1983, "barrel-hunts" ensued throughout Western Europe. After seven months the barrels were found in an unused slaughterhouse in northeastern France. Although not directly involved with the Italian explosion or connected with the responsible waste generator, the German Chemical Manufacturers Association (CMA) expressed its anger at this mismanaged waste. While most firms in America keep away from public statements on industrial accidents such as the recent Union Carbide releases and on the five-year fight against ocean incineration, the German CMA speaks out forcefully on this and related issues. Because the reputation of chemical manufacturers throughout Europe was tarnished, the German industry announced that it saw it as incumbent upon the industry itself to have prevented the catastrophe.[7] Perhaps this is nothing but good business sense. Yet the motivation for CMA's position was well outside any legal mandate and evidenced a position that went well beyond monetary strategies.

Of the many forces upgrading German waste handling today, the most colorful sparks still originate from the Green party. As the first and most successful party in an advanced industrial nation to base its programs on grassroots environmentalism, the Green Movement makes citizen activism its very core. Hamburg's Thea Bock, then chair of the region's decision-making body, noted that grassroots activism is the philosophical and actual foundation of this party.[8] This base continues to grow: the Greens secured 5.6 percent of the national vote in March 1983, and recent polls show that nearly 10 percent of the voters in 1985 support the Greens.[9] Since any political party receiving more than 5 percent of the votes receives representation in the national Bundestag, the Greens' dissenting voice is now a force to reckon with.

The Greens endorse environmental positions that are quite compatible with mainstream American environmentalists. They support reducing toxic waste problems at their source by first reforming the processes resulting in toxic wastes. They seek better controls on incinerators. They want to see recycling mandated by law.[10] Yet a difference resides in the Greens' activist tactics for reaching those ideals, as well as in their resolute refusal to engage in the conventional means of political compromise.

They speak of themselves as an "anti-party party,"[11] rejecting many routine aspects of political behavior that they view as corrupt. For instance, the Greens have rejected the concept of serving a full term. Once the term is half-over, a new set of representatives takes over. This is based on the principle that allowing a greater number of individuals to gain political experience prevents the representatives from becoming distanced from their grassroots support. The Greens also resist the growth of an elite of "professional party leaders." They claim that the perverse development of "the boss" is what has weakened the established German parties.

Serving a political term, however, does not replace the importance of direct action. This emphasis on direct action derives from the origin of the Green Movement. The 1960s marked the rise in Germany of massive political alienation, especially among its educated youth. This first wave surfaced in protest of the Vietnam War. But whereas youth protests weakened in most democracies during the 1970s, these groups remained vigilant throughout Germany. The political scientist Horst Mewes explained:

German environmentalism, ignored throughout the 1970s by the political establishment, arose as part of a broader, fundamental opposition to German politics and was based on the experience that apparently only direct citizen protest can persuade the political and economic experts to pay any attention to environmental deterioration.[12]

These direct citizen protests are most frequently organized by small local citizen groups, known as the *Bürgerinitiativen*.

By 1980 over 50,000 of these groups had formed. Mewes continued, "As spontaneously created protest groups, arising in the experience of everyday life, addressing the grievances to local government units . . . , these citizen groups proved to be effective."[13] By 1975 43 percent of the *Bürgerinitiativen* had taken actions on a full spectrum of environmental issues involving land-use planning and toxic contamination. In 1982 16 leading *Bürgerinitiativen* formed the BBU, an umbrella organization of environmental citizen groups that now consists of over 3,000 base groups.

Although most Greens agree that the party cannot and should not replace the grassroots base of the *Bürgerinitiativen,* there is an ongoing debate about whether the party should embrace all social movements. In addition, a persistent clash exists between factions of the party concerning working in coalition with established parties. The "fundamental oppositionists" refuse to join a party, claiming that the gesture itself is against Green principles, while the "political realists" claim that it is the only expedient manner for expressing Green values in policymaking.[14]

Although it is still too early to measure the feasibility of these proposals, their presence has already influenced the national debate. The Greens' mere existence on the political scene has also raised the environmental consciousness of the established political parties. The established parties fear a loss of voters in future elections.[15] As a result each party is trying to "outgreen" the other. The Social Democratic party (SPD), for instance, would like to pass legislation that assesses a waste-end tax based on the quantity of hazardous waste that must be managed. There is already such a system in existence for wastewater discharges into rivers.[16] In addition, the SPD now supports the passage of a freedom of information act in response to the government's current policy of withholding waste-monitoring results from the public. The Christian Democratic Union, however, does not support a waste-end tax on the production of hazardous waste; instead, it supports the continued enforcement of the wastewater discharge tax.[17]

Both established parties now endorse the printing of an "environmental price" on consumer goods, whereby labeling would give people more opportunity to make environmental choices. The consumer would receive information on the product that would indicate either how much of a hazardous substance had to be disposed of from the creation of the product, or what the product's relative hazard is compared to other products. The debate over the best approach for West Germany is still very much alive. The Christian Democratic party's environmental spokesperson found such a method philosophically agreeable because it leaves to the consumer the decision of whether or not to buy the more hazardous product.[18] This notion of equating environmental concerns with the purchasing power of individuals is doubly populist: first, it establishes a direct link between consumption and responsibility; second, it publicizes an objective gauge on a product's environmental impact, measured in toxic emissions per unit of resource used. This development could have widespread consequences for American consumers if the German trend is translated to concerned citizens in the United States.

In addition to these important party developments, Germany's many environmental groups continue to grow. In general, their aims and strategies are informally divided. The BBU, the umbrella organization for citizen groups mentioned earlier, emphasizes demonstrations joined with concerted efforts to lobby for legislative reform.[19] The BUND, described by its own staff as having "a lightly conservative touch,"[20] works for technological fixes on toxics as well as decentralized energy sources. Greenpeace, whose planning offices are often linked with a ground-floor shop that sells posters and pins on a central market or transport center, concentrates on direct action campaigns of international consequence.[21] The summer of 1985 marked Greenpeace's first major efforts inland: they now have a "multinational lab ship" testing toxic discharges into the Rhine. Germany also has the Oeko-Institut, founded to provide citizens with scientific expertise on nuclear power and other environmental issues. Like Greenpeace's campaign to inform the public, the institute now disseminates information on a wide range of toxics issues.[22] Each of the above groups is actively working on hazardous waste issues, with toxics reform campaigns clearly in the forefront.

Recently, the Oeko-Institut publicized plans for a major two-year project on toxics. While promising to analyze emerging problems, the project also will suggest corrective actions that must be taken—both in science and the legislatures. At the top of its list the institute asked for a freedom of information act. Moreover, wastes generated by consumer goods would be printed on labels similar to the listing of ingredients. The institute also supported a tax on manufacturers that directly reflected the environmental cost of products in terms of toxics generated per product.

The institute's most surprising request called for the creation of a voluntary fund to pay for testing and remedial work on old toxic dumpsites in instances where responsible parties cannot be found. When pressed as to why it seeks a voluntary arrangement, a representative said a voluntary fund would receive more support from the chemical industry than a mandated one. In fact, the institute expected a greater number of contributions if such an honors principle were established. Here the force of peer pressure again empowers a West German alternative to litigation.

Citing the German CMA's anger at the Seveso incident, an institute spokesperson felt that the chemical industry might want to contribute to this fund to prove its willingness to "clean up its own act." After the voluntary fund is established, the goal is to incorporate such an arrangement into law.[23] Confident that conceiving of voluntary funds was a sound first step, the institute was not to seek the most far-reaching legally binding solution first but instead was to approach the issue by a traditional appeal to conscience and industrial reputation.

According to an environmental attorney, citizen groups infrequently press their causes in court.[24] In addition to the problem of acquiring the data necessary to prove damage or injury to health or the environment, citizen groups never have standing to raise legal issues themselves; only individual members do.

Another feature of the legal system severely impedes formal court actions. The losing party pays the prevailing party's costs on all accounts. Thus citizen groups fear the financial penalty of their members undertaking difficult lawsuits.

This is not to say that litigation is not a tool used for environmental protection. There have been major legal victories, especially in the area of nuclear power. Nonetheless, German environmentalists do not turn to litigation with the same frequency or readiness as their American counterparts. In place of legal confrontation, the Germans depend on a tradition of mutual scrutiny that is both informal and forcefully intense.

SUBSIDIZED DISSENT: THE DUTCH USE OF ENVIRONMENTAL CRITIQUES

With hundreds of environmental organizations directing local, provincial, and national actions, the Dutch environmental scene constitutes a rich tapestry of intricately interwoven threads. With fewer than 13 million residents, Holland has one of Western Europe's strongest environmental movements. It is often heralded as the most progressive in tactics and product.[25] Perhaps part of the strength is derived from the size of the country. The distance psychologically and physically between the ordinary citizen and the national government is much shorter than in most other countries.

One national environmental group, Stichting Natuur en Milieu, openly discusses technical and policy issues with government and citizen groups alike. Often it reports its results to both sectors at the same meeting. The group has representation on all important governmental commissions and is an expected player in any toxics debate.

The Dutch environmental movement plays an integral role in environmental policymaking and is treated, financially, as a partner. Many environmental groups receive subsidies from the local, provincial, and national governments. In providing the groups with funding, the government deliberately frees these organizations from some of the fundraising burden that often consumes sizeable amounts of staff time for American environmental groups. This subsidization is not extended just to environmental groups. The Dutch governments also fund many civic, political, and artistic endeavors.

Environmental attorneys also encourage clients to seek governmental funding to pursue common goals based on the principle that their interests represent the long-term public interests. On occasion the government has agreed to subsidize even these citizens' legal claims. An attorney working for the Ministry of Housing, Physical Planning, and the Environment explained that Dutch authorities "use these monies as early warning signals. If an environmental group requests increased subsidies on a certain issue, we check to see if we're giving the problem enough attention within government."[26]

Asked whether the government funding might sway Dutch environmentalists to seek more comfortable relations with the government, individual environmentalists said that the funding may affect the style of the groups, but most believed that it did not influence the positions espoused by the groups. A retired chemist was the single outspoken exception. Claiming that subsidization has institutionalized the movement, he believes that Dutch scientists too often have been replaced by social scientists more likely to buckle under their own perceived threat of losing government funding. He also felt that the subsidized groups are now less aggressive spending too little time assisting local citizen groups. Despite this single respected critique, most respondents showed the Netherlands to be a nation quite proud of its policy of subsidization.[27]

Since much of Holland sits below the water table, fill is often required before construction can take place on the land. For over 40 years solid wastes containing hazardous chemicals were often utilized as these fill materials. A 1980 government survey estimated that there are now nearly 5,000 waste sites that are leaking from this past practice alone. According to Hans Erasmus, director of the Ministry of Housing, Physical Planning, and the Environment, "More than 1000 locations require further study, with 750 situated in residential or water collection areas ....The clean-up will cost billions of Dutch guilders [over $300 million]."[28] Many local citizen groups have formed over the discovery of these local problems. One environmental group, Nederland Gifvrij (Poison-free Netherlands), functions as an umbrella organization that puts citizen groups in touch with each other to exchange information and share notes on tactics for solving their mutual problems. Its goal is not dissemination of technical advice, but rather efficient networking among citizens.

The Dutch people are highly alert to toxic waste issues. Perhaps much of this awareness is the result of the toxic problems at Lekkerkerk. The story of Holland's Love Canal deserves note, showing in microcosm how the Dutch people seek to resolve environmental disputes.[29]

The small village of Lekkerkerk was expanding with construction of new housing in the 1970s. Before construction was even completed, the water mains corroded and had to be replaced. Farmers in the area recalled spontaneous fires erupting during this period. Yet no one feared toxic contamination. It took a set of new residents, made slowly aware of mutual problems by the persistence of their air, water, and soil problems, to form a citizen group.

With no law to provide public access to planning information, a few Lek-kerkerk citizens surreptitiously made copies of documents indicating that their local government allowed the dumping of chemical wastes as fill material without a chemical waste permit. What had been simmering was now brought to a boil. With evidence that 1,650 drums of paint and industrial waste were dumped in their neighborhood, the citizens brought the issue to the press. Meanwhile, the local government issued a retroactive disposal permit without public hearing or comment. Enraged at this atypical event, the citizens received the active support of 130 of the 300 families in town.

Acting on the belief that the local officials had aided and abetted the dumpers, the group bypassed local and provincial authorities. Instead, they pleaded directly before the national government and the queen. The response was enormous and swift. The citizen group and others believe in retrospect that the government "panicked."[30] Within a six-month period culminating early in 1980, the government answered the most critical of the citizens' requests. Bottled drinking water immediately arrived. Residents were rapidly evacuated so that the contaminated fill could be excavated. The government then bought the contaminated homes at their fair market value before the threat of contamination devalued the area. Even renters in surrounding mobile homes were evacuated. Perhaps most astonishingly, monies were provided for child care during the evacuation time as well as funds to hire attorneys to assist the citizens. In sharp contrast to the gruesome account of initial governmental inertia found in Michael Brown's *Laying Waste: The Poisoning of America by Toxic Chemicals,* the tale of Lek-kerkerk embodies the benefits of a rapid response to citizen unrest on environmental matters.

The leader of the citizen group spoke of how hard they had fought for every one of their concerns. And indeed they had, for there had been no precedent they could follow. Yet a six-month time frame in this country for a toxic waste problem of such proportion remains unthinkable. The Lekkerkerk citizen group was in contact with the Love Canal Homeowners Association, since they shared similar struggles at that time. Later, the leader of the Dutch citizen group was terribly saddened after watching a film of Love Canal to see how much more the residents there had to struggle and suffer.

Perhaps most intriguing about the Dutch governmental reaction is that in a panic the government preferred to overcompensate.[31] Lacking a legislative mandate for remedying its unprecedented contamination problem, the government acted out of a sense of its duty rather than its law. The Lekkerkerk episode illustrates the considerable reliance Dutch society places on its government in its pursuit of consensus and resolution.

In some Dutch towns local citizen groups have chosen to bypass the plethora of environmental organizations.[32] On the surface, this calls into question the usefulness of the subsidized environmental groups. Dutch citizen activists, however, avidly defend the environmental groups. For them, the existence of the environmental organizations provides a backdrop against which the citizen groups may be more easily understood and supported. The Dutch citizen activist and environmental groups are perceived as symbiotic despite their separate planning and purposes.

Dutch environmentalists frequently use symbolic acts for educational purposes. From October 2 to 7, 1983, for instance, an International Water Tribunal was held in Rotterdam. Environmentalists from the Netherlands, Denmark, West Germany, Great Britain, Belgium, and Switzerland were important participants. The tribunal was convened to hear 19 environmental charges against 39 large companies. This forum, however, was merely symbolic. It commanded no power of law. Yet environmentalists spent two years gathering 10,000 water samples, the results of which they presented dramatically before the press and attendants.

In Holland, it should be noted, the government actually contributed money directly to the tribunal organizers. Governments and industry were offered time before the tribunal, but neither chose to participate. Many citizen groups were skeptical of such actions, so they allowed the environmental groups to participate while they concentrated on their local efforts. The environmental groups that participated felt overwhelmingly that the tribunal had a great deal of impact.[33] The press covered the hearings daily. The participants assembled an 800–page reference book for future use. Environmentalists, not bound by formal rules of evidence, examined the legal and ethical consequences of the notion of a "permission to pollute." In one instance, at the conclusion of the presentation of a charge, the provincial government instituted its own lawsuit. As an indication of the neutrality of the tribunal, not all environmental groups won. The jury was considered "moderate," and not "green."[34]

This is but another example of relying on mechanisms that in and of themselves do not set in motion specific legal solutions. The tribunal existed outside of any legal structure. Yet many European environmentalists chose to spend much time and money to carry it off. This may be because of the barriers to legal action that exist in Holland.

As in Germany, citizen groups fear the financial risk in undertaking a lawsuit because of the taxing of costs to the losing party. Under the temporary statute on remedial action for soil and groundwater pollution, citizens may object to proposed remedial action. However, unless the provincial government fails to place a site on the list, citizens may not appeal. In addition, as one environmental attorney explained, Holland is simply not a litigious culture, and even if barriers were removed, there would not be much litigation. Nevertheless, Holland recently enacted a Freedom of Information Act that should assist access to the courts. From legal and citizens' perspectives there are some remaining problems with the implementation of this law, but all agree that it is a quite useful development. Asked if a Dutch Green party was needed, those interviewed by the authors often felt that it might be helpful. But many questioned whether it was as necessary as in West Germany because of the manifest strength of the Dutch environmental groups.[35]

AN INSULATED TRADITION: CITIZEN INDIFFERENCE IN BRITAIN

The cornerstone of Great Britain's toxic waste plan is co-disposal: the deliberate mixing of hazardous wastes wih conventional municipal wastes in permeable landfills. Based on the belief that the leaching of toxic chemicals will change the wastes into non-toxic substances over time by dilution and biological degradation, co-disposal is at odds with positions held by EC directives, the United States Environmental Protection Agency, and environmental officials in West Germany, Canada, and Australia.[36] Even cyanide, PCBs, cadmium, mercury, and arsenic (in small concentrations) are considered suitable for co-disposal in England.[37]

The goals of co-disposal are diametrically opposed to those of secure landfilling.[38] While daily cover is required to reduce infiltration in secure landfills, co-disposal seeks as permeable a surface as possible. The former confines waste to the smallest area, whereas the latter seeks the largest area to maximize evaporation. Nonetheless, Britain, dependent mainly on surface water for drinking, has not experienced to date any major environmental catastrophe with toxics. Thus, while Lekkerkerk and Love Canal have forced Dutch and American planners to confront the inherent problems with programs based on the capacity of the land to contain or transform hazardous by-products, British authorities remain articulately smug over their dependence on land-based systems. David Mills, the principal chemist for the Land Waste Division of the Department of the Environment, noted in print: "It is fair to say that the United Kingdom has achieved high standards of environmental protection and public health related to waste management long before most of the other European nations."[39]

Perhaps as a result of this tradition of confidence, there are few environmental groups in England concentrating on toxic waste issues. Even less are citizen activists or citizen groups struggling with the issue. In sharp contrast to the Dutch, who appear to specialize in self-critical cross-examinations, British officials appear proud in a broad, open manner. A volunteer chemist, concentrating on toxic waste matters for an environmental group, did express some reservations about co-disposal. He felt that policies regulating domestic wastes should not allow careless landfill practices, since much of the bulk of the waste could be recycled. He also supported process modifications as well as increased toxics recycling to reduce toxic wastes.[40] Yet overall, the people of Britain remain consistently indifferent to the issues surrounding land disposal.

The official British technical data supporting co-disposal was presented to an American hydrologist. Will it work? "Yes, if you've got a quarter of a million years."[41] The response encouraged a further survey of Britain's official position. Today, David Mills's confidence in the resilience of British soils is no longer universally held. Yet the suspicions are arising from government enquiries, not citizen activism.

The House of Lords' Select Committee on Science and Technology prepared a report on hazardous waste disposal in 1981 that expresses some concern about British disposal methods: "Too many people are comforted by the belief that because nothing much has gone wrong so far, nothing is likely to go wrong in the future."[42] However, of the 641–page transcript of written and oral evidence submitted to the Select Committee, only six pages were submitted by environmental groups and none by citizen activists. This profound lack of citizen activity illustrates an important point made by Lord Ashby:

> To set the chain reaction going is often the hardest task in social reform. This raises an ethical problem of some importance. Is it morally defensible to use shock tactics, to exaggerate, to distort the facts or color them with emotive words, or to slant the TV camera in order to excite the public conscience? My experience leads me reluctantly to believe that in the present social climate some dramatization is necessary.[43]

As a forerunner of much of America's civic and legal traditions, a strong chain exists between public opinion and government response in Britain. Yet in its present social climate, citizen groups have not begun to put pressure on this link. The dramatization of the problem, evident in the activities of Lois Gibbs's group at Love Canal and Ben van Veen's at Lekkerkerk, is simply missing in Britain. Without this dramatization, co-disposal remains the single dominant means of waste management.

Since the authority to regulate the operation of a co-disposal site is vested in the County Waste Disposal Authorities, citizens have no standing to sue to stop such dumpsites in England. Citizens can request and apply pressure through the press for the responsible agency to hold a public inquiry, but in general this lack of court standing serves as a severe disincentive to toxics activism. At a public inquiry, opposing parties present evidence, rally experts, and circulate testimony to both government and the press. As a forum for the airing of views, the procedure is similar to American administrative hearings. But since the enquiry is not required by law, the conclusions do not mandate action.

Despite these legal restrictions, British groups have used public inquiries forcefully in disputes over nuclear reactors. Friends of the Earth invested over 70,000 pounds (over $40,000) in their Windscale reactor inquiry, rattling the project considerably.[44] Another large-scale enquiry underway in 1984 concerned the Sizewell nuclear reactor. Friends of the Earth expected to spend even more time and money on Sizewell in an effort to stimulate a large wake of protest. Why similar tactics, which can delay projects dramatically or change policy directly, are not used on large co-disposal proposals remains unanswered.

Citizens also might check or openly question waste management decisions by filing a complaint with their ombudsman. Targeted at a specific government official, the complaint focuses on whether a certain person—not a policy or rule—is executing his or her job properly. The ombudsman then decides if the allegation of maladministration is valid. Thus the technique works best as a lever for responsible decision making yet remains less effective as a tool for questioning an entire waste management strategy. But since such complaints are taken very seriously in Britain,[45] the ombudsman's review might serve citizens by focusing a debate on toxic waste issues. To date, this approach also remains unused in England.

In summary, two devices that exhibit British traditions to manage disputes through informal and non-legal means remain unexercised on toxic disposal issues. Since both these devices—the public enquiry and the complaint to an ombudsman—have been used by citizens in other environmental contexts, the absence of their use on toxics issues is a clear confirmation of Ashby's further claim:

Public opinion has to be raised to a temperature that stimulates political action. In the second stage the hazards have to be examined objectively, to find out how genuine and how dangerous it is and just what is at risk. In the third stage this objective information has to be combined with the pressure of advocacy and with subjective judgments to produce a formula for a political decision.[46]

Britain cannot reach Ashby's second stage of a public enquiry, let alone the further stage of formal complaints against land disposal, until the temperature of the public is raised considerably. For the nation that established John Locke's conception of the sanctity of private property, this does not occur by government reports alone.

THE POWER TO FORGET: ITALIAN CITIZENS AFTER SEVESO

On July 10, 1976, in a chemical plant owned by a subsidiary of the Swiss pharmaceutical firm of Hoffman-LaRoche, a switch was left unattended before the workers went home for the weekend. Over the weekend, large quantities of 2,3,7,8–TCDD—the most toxic form of dioxin known—were inadvertently produced. The cover of the reaction vessel exploded and a cloud of dioxin hung over the town. Within hours, the eyes of the world were upon Seveso, a small rural Italian town only twenty kilometers from Milan's two million residents. Within two days, hundreds of small animals and birds were dead.[45] The long-term effects on Seveso's residents are uncertain, although no apparent serious health effects have ensued according to Italian officials.

Today there are no citizen activists or environmental groups found in Seveso. Moreover, in Milan not one environmental group works on toxic chemical issues. During the crisis the state of Lombardy had appointed a commission to manage the accident. Representatives from this Ufficio Speciale recall that citizens did request information in the months following the explosion, so newsletters with updates were provided. But within a year the affected public grew utterly disinterested.[47]

Within the greater Milan area, Seveso is not remembered as a monumental crisis in world environmental history. Instead, in an ironic twist, it is best known for the role Seveso citizens played in the legislation on abortion.[48] During the first half of 1976, a public debate raged throughout the Italian press on the question of abortion. Press attention, soon after the accident, centered on pregnant women in Seveso. Shouldn't Seveso's women, exposed to dioxin, have the option of a legal abortion? In early August, the Ministry of Justice ruled that a 1975 decision, which allowed abortions in those cases where the psychological stability of the mother was threatened, was deemed applicable to the women of Seveso. Thus, if a woman claimed that the dioxin exposure created a psychological threat, she was entitled to a legal abortion without the need to prove physical contamination. In fact, 26 women received abortions in the middle of August— marking the first set of legal abortions performed in Italy.[49] Yet no legislation has been passed to manage future industrial accidents in Italy.

In seeking to understand this absence of response in Italy, one must acknowl- edge the inherent difficulties government faces in responding to low-probability accidents that have alarming, high-visibility consequences. A psychologist work- ing in a neighborhood clinic near Seveso in July 1976 said that denial of a catastrophe is one of two dominant reactions (the other one being inarticulate hysteria).[50] For the most part, denial by residents was the overriding reaction in Seveso from the start of the incident. Most of the people who were evacuated returned home in a short time. There is some belief that the socioeconomic background of those affected enhanced this reaction. Seveso was inhabited by many who had left the south for better economic opportunity. They had begun life in Seveso as squatters, building up their shacks and huts into comfortable houses. Thus money became one of the most important decision criteria for the affected victims.

The national and regional governments filed suit against the Hoffman-LaRoche subsidiary shortly after the incident. A court panel of experts unanimously found the subsidiary liable for the accident. The company chose to settle, paying $128.9 million for administrative costs to the governments for cleanup, reclamation, and health studies.

A geologist now working on sociological aspects of environmental issues offered some explanations. Through a comparison of major European daily newspapers, he has concluded that Italians have a consistently low interest in environmental affairs.[51] He connected this indifference with the Italians' histor- ical sense of themselves. He believes that the accident was inconsistent with their mission to maintain a sense of the grandeur of their past, during the period when Italians commanded international cultural and financial attention. As a result, preservation of the natural environment is still not as important a part of Italian heritage as art and historic preservation.

AFTERWORD

Environmental decision making, at its best, is more than science. It is a process by which a society comes to know its environmental stresses, and then structures an economic response to the challenge in a manner both humanistic and pragmatic. This may be the ultimate lesson derived from the above four test cases: to learn technical competencies is not enough. The capacities to weigh the risks of a pollutant and use or design treatment equipment clearly represent needed and valued skills in West Germany, Holland, England, and Italy. Yet the preceding studies suggest that a second set of prerequisites is also in demand.

Toxic wastes, by their intrusive properties, demand environmental decision makers who can comprehend conflicting perspectives, discern reliable evidence from partial information, and then make sound judgments. Citizen and environmental groups, as shown in Holland and West Germany, can play vital and influential roles in assisting these decision makers. To underestimate the value of citizen contributions is to shortchange the attempted environmental reform.

While most American groups still restrict their inputs along technical and legal lines, the Dutch and German groups have begun to expand the decision model itself. Whereas the old model was based on a politics exclusively determined by experts, the Dutch and German groups seek a new politics that is far more consultative and participatory. Moreover, with their willingness to use direct action, their new decision model actually emphasizes the use of citizens to infuse environmental foresight into planning. Finally, whereas the old politics tied most of its environmental disputes into questions concerning the management of the economy (which thereby required most proposed solutions to emphasize almost exclusively market controls), the new environmental groups of Holland and West Germany function in a more value-laden and people-oriented world.

These European groups express a generalized compassion for other species, other peoples, and subsequent generations. They seek changes in the behavior of citizens, wanting to link purchasing interests with a knowledge of toxic consequences. In short, they want to retire the old model of decision making. In its place they seek consensus-building groups that pursue social good and environmental protection with a strategically lessened reliance on legal tactics and direct confrontation. While it would be simple-minded to claim these European innovations answer all of America's conflict-ridden needs, it is simply neglectful to ignore the intelligence of these pathways for environmental conflict resolution much longer.

NOTES

1. Mrs. Schimmelpfennig, Aktion Besorgter Büerger (ABB), interview, Stolberg, February 16, 1984.

2. Bernward Boden, Bürger Initiative Wohnen u. Umwelt, interview, Cologne, February 10, 1984.

3. Guenter Glatz, BI Wilhelmsburg/Neuland, interview, Hamburg, March 28, 1984.

4. Dr. Dieter Schottelius, Verband der Chemischen Industrie e.V., interview, Frankfurt, March 30, 1984.

5. Mr. Szelinski, Umweltbundesamt, interview, Berlin, March 26, 1984.

6. Carl-Otto Zubiller, Hessen Ministry, interview, Wiesbaden, August 1986.

7. Dr. Dieter Schottelius, Verband der Chemischen Industrie e.V., Frankfurt, March 30, 1984; Rainer Griesshammer, Oeko-Institut, Freiburg, interview, March 14, 1984.

8. Thea Bock, Chairperson of GAL Fraktion, Hamburg, interview, February 14, 1984.

9. For a fuller account of recent developments within the Green Party, see Horst Mewes, "The Green Party Comes of Age," *Environment* 27, no. 5 (June 1985): 13–17, 33–39.

10. Barbara Zeschmar, adviser to Green Party on Environmental Matters, interview, Bonn, March 7, 1984. For the early history of the Greens see Horst Mewes, "The West German Green Party," *New German Critique* 28 (Winter 1983). For a readable reflection on the possible applications of Green Politics to other nations see Fritjof Capra and Charlene Spretnak, *Green Politics: The Global Promise* (New York: E. P. Dutton, 1984).

11. Rhea Thoenges, Green state representative, interview, Kassel, March 18, 1984.

12. Horst Mewes, interview.

13. Ibid.

14. Thoenges, interview.

15. Ibid.

16. Mewes, "The West German Green Party," p. 15.

17. Otfried Klein, SPD advisor on environmental matters, Bonn, and Rainer Gross, ODU/CSU advisor on environmental matters, interview, Bonn, March 8, 1984.

18. Klein, interview.

19. Gerd Billen, Bundesverbandes der Buergerinitiativen Umweltschutz, (BBU), interview, Bonn, March 13, 1984.

20. Lutz Ribbe, Bund fur Umwelt u. Naturschutz Deutschland (BUND), interview, March 12, 1984.

21. Monika Griefman, Greenpeace, interview, Hamburg, February 13, 1984.

22. Peter von Gizycki and Rainer Griesshammer, interview, Freiburg, March 14, 1984.

23. Siegfried de Witt, attorney, interview, March 15, 1984.

24. Konrad von Moltke and Nico Visser, *Die Rolle der Umweltschutz-verbaende im politischen Entscheidungsprozess der Niederlande* (1982), p. 89.

25. Renee Buisman, Nederland Gifvrij, interview, Utrecht, February 20, 1984. Ryk van der Hoek and Ellen van der Linden, Burgerkomitee/Vereniging tot behond van Waterland, interview, Laan, February 20, 1984.

26. Meeting at the Ministry of Housing, Physical Planning and the Environment, July 1985.

27. Dr. J. W. Copius Peereboom, interview, Amstelveen, February 22, 1984.

28. Hans Erasmus, "Industrial Hazardous Waste Management in the Netherlands," in UNEP's special issue (no.4) *Industrial Hazardous Waste Management* (1983), p. 25.

29. Story as told by citizen group leader Ben van Veen, Krimpen A/D Yssel, interview, February 24, 1984.

30. Ibid.; Hugo Pennarts, attorney, interview, Rotterdam, February 23, 1984.

31. von Moltek and Visser, *Die Rolle der Umweltschutz-verbaende im politischen Entscheidungsprozess der Niederlande,* p. 8.

32. van der Hoek and van der Linden, interview.

33. Liesbeth Beekhuis, Contact Milieubescherming Noord-Holland, interview, Zaandam, March 23, 1984.

34. Ibid.

35. Pennarts, interview.

36. Brian Copes, House of Lords Select Committee on Science and Technology, *Hazardous Waste Disposal,* vol. 2 (1981), pp. 135–70.

37. Department of the Environment, Waste Management Paper no. 4, *The Licensing of Waste Disposal Sites* (1976), p. 31.

38. House of Lords, Select Committee on Science and Technology, *Hazardous Waste Disposal,* vol. 1 (1981), p. 120.

39. David Mills, principal chemist, Land Waste Division, Department of the Environment, "National Hazardous Waste Management Practices in Western Europe," *Chemistry and Industry* (June 6, 1983): 423.

40. Brian Price, environmental pollution consultant, interview, Weston-super-Mare, February 28, 1984.

41. Dr. John Dowd, hydrologist, University of Georgia, interview, May 3, 1984.

42. House of Lords, *Hazardous Waste Disposal,* vol. 1, p. 2.

43. Lord Eric Ashby, *Reconciling Man with His Environment,* (London, Oxford University Press, 1978), pp. 21–22.

44. Brian Wynne, *Rationality and Ritual: The Windscale Inquiry and Nuclear Decisions in Britain* (Bucks, Eng.: British Society for the History of Science, 1982).

45. Colin Palmer, waste disposal engineer, Suffolk County Council, interview, Ipswich, March 2, 1984.

46. Ibid.

47. Dr. V. Silano, *Case Study: Accidental Release of 2, 3, 7, 8 Tetrachlorodibenzo-p. Dioxin (TCDD) at Seveso, Italy* (Copenhagen: World Health Organization, 1981), p. 167.

48. Dr. Umberto Fortunati, Ufficio Speciale, interview, Seveso, March 22–23, 1984.

49. Ibid.

50. Michael Reich, *Toxic Politics* (Ph.D. diss., Yale University, 1982), pp. 91, 98, 99.

51. Dr. Leda Dacquisto, psychologist, interview, Milan, March 22, 1984.

10

Establishing Collection Systems: Benefits of the Infrastructure Approach ___

BRUCE PIASECKI AND STUART MESSINGER

Generators of small quantities of hazardous waste face unique management problems that are only beginning to be publicly recognized. On November 9, 1984, the U.S. Congress moved to redefine small quantity generators, in accordance with the Resource Conservation and Recovery Act (RCRA), as those firms that produce 100 kilograms per month. Until that date, the small quantity exemption applied to any firm that produced less than 1,000 kilograms per month; these firms could discard their waste as if it were non-hazardous municipal waste. Thus, in a single gesture, Congress increased the number of regulated generators by over 700,000 firms. Many of these smaller generators, such as some gas stations, dry cleaners, and paint, leather, and electroplating shops, lack a professional environmental staff and will have to make major adjustments to comply with this refinement of the law.

Small firms responding to the new regulations will get their first experience with complicated waste management regulations. In addition, they will be faced with new expenses in the form of high waste disposal fees. Small generators who are not familiar with the regulations, or those wishing to avoid the added expense of proper disposal, may continue to send hazardous wastes to a local landfill. Other generators may resort to illegal disposal along roadsides or pouring

The authors wish to acknowledge the assistance of Dana Duxbury of Tufts University's Center for Environmental Management for her timely assistance in assembling Appendices A and B to this chapter.

their wastes down the drain. Clearly these industries will need assistance if proper management is to be ensured.

Household toxic waste disposal is a related problem. Florida and Massachusetts have taken the lead in providing programs to collect and dispose of household hazardous wastes; many other states and communities have followed suit. These American programs are still in their infancy often serving as educational tools rather than an effective means for managing significant quantities of wastes.

The United States is well behind some European countries in dealing with small generator and household toxic wastes. (Table 10.1 gives a summary of recent American developments.) In Europe, advanced and comprehensive collection and treatment systems for small quantities of hazardous waste have been in existence for over ten years. What follows compares the best of America's programs with the added benefits of the European infrastructure approach. Once again our goal is to identify the optimal mix of both traditions.

THE NEED FOR REGIONAL COLLECTION PROGRAMS

What are household toxic wastes? Households are not regulated generators of hazardous waste. They are specifically exempted under RCRA. In addition, the concentrations of toxic waste produced in households nearly always fall below the commercial and industrial grade concentrations required for reuse or recycling. This incapacitating restriction requires increased government involvement. However, wide varieties of potentially toxic substances are disposed of by the average American household. Over 90 percent of all households in the United States use pesticides and disinfectants.[1] A waste characterization study in Los Angeles indicated that household cleaning and automotive products together make up 70 percent of the total household hazardous waste containers found in residential solid waste.[2] Table 10.2 presents this data.

Of interest in this table is the fact that 92 percent of all the containers were empty. This suggests that "targeted" collection programs might work by removing the most ubiquitous and problematic wastes first. A prime example might be used motor oil, which is collected and recycled in many communities as assessed in Chapter 6. Other candidates for concentrated collection programs might include unused pesticides or batteries.

Since waste disposal is now practiced in most of the United States, container residues and partially filled containers will still end up in the same place as the rest of the community's waste—generally the town dump. If the dump lacks liners and leachate collection systems and is poorly sited ("wastelands" such as wetlands are popular spots for the older generation of landfills), then groundwater contamination is inevitable. Even modern, well-designed sanitary landfills, as documented in Piasecki's *Beyond Dumping* (1984), frequently experience contamination. This is partly due to the highly reactive and corrosive nature of many wastes in the acidic landfill environment. The fact that all landfills produce a highly polluted leachate has been well documented.[3]

Table 10.1
Examples of Household and Small Business Hazardous Waste Disposal Programs

Location (date)[b]	Collection site	Amount of waste collected[a]	Number of participants	Project cost	Destination of waste
Anchorage, AK (1982)	Municipal shredding facility	1,000 lbs plus 35 barrels waste oil	48 households 41 businesses	$50,000 plus time and equipment donations	Stored at sealand van; shipped to Oregon for burial
Palo Alto, CA (1983-84)	Recycling center	Fall:30 55 gal. drums; spring:55, 55-gal. drums	150 households	$3,900 fall; $6,100 spring (included disposal cost)	Landfill
Redlands, CA (1984)	Public site	175 gals. liquid; 76 lbs. solid	30 households	N/A	Landfill
Sacramento, CA (1982,84)	1982:hazardous waste transfer station; 1984: shopping mall, school and state parking lots	1982:54 drums of hazardous wastes and 2,400 lbs. of waste oil for recycling; 1984:165 drums	1982:250 households 1984:900 households	1982:$14,000 1984:$25,000 plus in-kind contributions (both years)	1982 and 1984: Class I hazardous waste disposal site; 1984:latex paint reused
San Diego, CA (1984)	Licensed hazardous waste transfer facility	13,626 lbs. in 5,057 containers	202 pick-ups, 88 people went to the collection site	$57,000:$47,000 education materials, logistics, etc., for one year; $10,000 for hauling	Class I landfill, motor oil recycled

Table 10.1 *Continued*

Location (date)[b]	Collection site	Amount of waste collected[a]	Number of participants	Project cost	Destination of waste
Woodland, CA (1984–85)	Public site	33,55-gal. barrels plus 100 gals. motor oil	106 households	$10,900	Class I hazardous waste disposal site
Florida (1984)	2 trailers moved around to easily accessible sites	250 tons of waste	50 schools, 86 government agencies, 2,513 households, 277 businesses	$550,000 for May and June (includes public education)	Class I landfill; paint products recycled; some waste incinerated
Barnstable, MA (1983)	7 regional sites at schools and supermarkets	8,000 gals in bulk plus 144 gals. of waste oil	500-plus households	$13,000:$2,000 education, $11,000 operations	Solvent recovery, high-temperature incineration, landfill
Lexington, MA (1982–83)	Public works facility	86,55 gal. drums	316 households	$1,800 for 1982 projects; $2,400 for 1983 project	Latex paint recycled, landfill, solvent recovery, high-temperature incineration

Seattle, WA (1982)	Fire dept.	Pesticides: 90 lbs. dry & 6 gals. liquid; solvents:3 qts.; oil:40 gals.	65 households	Pilot collection cost about $2,000 (part of a $50,000 project)	Recycled solvents; other waste landfilled
Madison, WI (1984)	School & service garage	2,872 lbs.	325 households	$12,000	Landfill
Victoria, British Columbia	Regional waste facilities	15,312 gals. PCB waste, 43,296 gals. other waste	Thousands over past six years	$615,000 (C) capital cost; $70,000 (C) operating cost	Storage facility and Class I landfill
Toronto, Ontario (1983)	Landfill 60 miles west of Toronto	3,477 containers (volume/weight unknown)	200 households	$17,045	Oil recycled; other waste landfilled

(a) Estimates may vary depending on methods used to calculate weights
(b) Indicates year(s) program has operated

This table was adapted from information developed by Golden Empire Health Planning Center of Sacramento, California, and K. P. Lindstrom and Associates; G. Purlin et al., *Household Hazardous Waste: Solving the Disposal Dilemma*, Golden Empire Planning Center, 1984.

Table 10.2
Characterization of Hazardous Residential Solid Waste from Los Angeles

Product	Total # of Containers Found	% of Total Sample	Total # of Containers Empty	% of Empty Containers
Automotive Products	618	30%	596	31%
Personal Items	338	16%	314	17%
Household and Cleaning Products	823	40%	779	41%
Insecticides and Herbicides	51	3%	43	2%
Paints and Allied Products	155	7%	103	6%
Other	71	4%	62	3%
TOTALS	2,057	100%	1,897	100%

Sources: S. Ridgely and D. Galvin, "Summary Report of the Household Hazardous Waste Disposal Project Metro Toxicant Program 1," Metropolitan Seattle Service District, 1982.

Attempts to trace specific leachate components to specific household wastes are underway but have not been reported. State environmental agencies generally require that leachate be tested for such parameters as total dissolved solids, chemical oxygen demand, chlorides, sulfates, phosphates, various metals, and the like. Such testing fails to identify the causal components, yet a more detailed testing protocol would be exorbitantly expensive for most communities. Thus the exact extent to which household toxic waste contributes to leachate toxicity is not precisely known.

Some householders pour wastes down the drain or toilet (thus emptying the container). As a result, pipes corrode, fumes are released, and the product may cause sewer and septic system malfunctions, eventually contaminating ground and surface water. Contamination in residential wastewater is the result of improperly disposed soaps, detergents and bleaches.[4] Organic pollutants, unlike bleaches and similar household products, are also found in residential wastewater.

Information that might trace these contaminants on some occasions back to individual households does not yet exist. Still, the possibility of citizen self-contamination cannot be ruled out in most inquiries.[5]

To complicate the issue further, some householders, particularly in rural areas, dispose of their waste by dumping it along roadsides. Other farmers maintain private dumps for their wastes. Old farm foundations are often filled with cans, bottles, and raw garbage. The types of toxic materials found in the modern waste stream makes this type of "pit" burial the least desirable form of disposal, since it creates the potential for environmental pollution from a wide range of materials. Moreover, disposal in streams or down the sewer (via drains and toilets), which happens in various small industries, is extremely undesirable and potentially damaging to public health and the environment.

Perhaps more crucial than household toxic waste dumping are the problems faced by the small commercial or industrial generator of hazardous waste. When the EPA first promulgated rules in 1980, it designated the status of "small quantity generator" to individuals or firms generating less than 1,000 kilograms of hazardous waste in a month (1 kilogram for certain acutely toxic wastes). As noted in the introduction, small quantity generators were thereby exempted from rules governing larger generators and were permitted to dispose of their wastes in municipal solid waste landfills. The rationale behind this ruling was threefold. It reflected Congress's desire to avoid unduly burdening the small businessperson with excessive regulation. Moreover, it spared the anticipated administrative costs and effort necessary to regulate so many generators. Finally, the rule was put forth in the belief that it would cover most of the waste produced by small businesses.[6]

The EPA originally estimated that of the 760,000 producers of hazardous waste, less than 10 percent produced 99 percent of all hazardous waste produced. Thus, the EPA exempted those producing less than 1,000 kilograms per month, concentrating on the medium and large generators. The Congressional Office of Technology Assessment later reviewed this estimate and concluded that 5 percent of the nation's waste stream was actually being exempted by the 1,000-kilogram cutoff.[7]

Subsequent experience indicates that hundreds of thousands of tons of hazardous waste enter solid waste landfills annually. This fact, evidenced by the growing number of municipal solid waste landfills on the federal Superfund list, prompted the realization that stricter laws were required to upgrade the management practices of small generators. In response, the RCRA reauthorization bill of 1984 required the EPA to develop rules regulating small generators at the 100-kilogram-per-month level and stricter standards for solid waste landfill design. States such as California (which has no small generator exemption) and New York have moved to lower the cutoff further, particularly with regard to the most problematic wastes. Over 700,000 of the newly regulated generators are now small businesses. As noted earlier many still need assistance to cope with the stringent management requirements of the new regulatory environment.

The problem is similar to that facing farmers early in the century, when agricultural techniques were clearly failing from overuse and inexpert soil management, but thousands of farmers lacked the information they needed to upgrade their equipment and improve the handling of their land.[8] The result was the creation of the Agricultural Extension Program, which put county extension agents into the field to educate farmers about the new techniques. A similar extension program initiated by federal or state-based legislation could benefit the small generator immensely. A pilot hazardous waste extension service is already underway at the Georgia Institute of Technology. But we must look to Europe (see next section of this chapter) for the fullest examples of an inclusive infrastructure.

The problems that small generators face have also been recognized in New York State. The New York State Governor's Hazardous Waste Treatment Facilities Task Force noted that most small generators are not familiar with state and federal regulations governing waste management. Moreover, they lack technical expertise in waste management and are generally unaware of alternatives. Frequently suspicious of regulatory agencies, small generators are reluctant to contact them for advice. Most importantly, they lack the resources to do more than the minimum required by regulation. In the face of these impediments, the task force recommended that an educational program be initiated for the small generator.[9] About $150,000 has recently been budgeted in New York for this purpose.

The pilot program mentioned above could serve as a model for a federal program to address small generators' needs, such as those identified by the New York task force. The pilot program is confidential, with no threat of EPA enforcement or involvement. It suggests that confidential consultation, customized services, and a technical and managerial process that enables the small firm to legally and effectively handle its own waste can lead to greatly improved management practices. Such efforts only need leadership and coordination by state or federal governments. An awareness of the European models that precede it also helps.

THE EUROPEAN PRECEDENT

The most advanced waste collection and treatment system in the world is found in Denmark. By law, each municipality must provide facilities for the collection of household hazardous waste. Denmark, a country of 5.1 million people and roughly twice the size of Massachusetts, has a network of over 300 collection stations, one located in every major community in the nation. In addition, mobile collection stations are used for "cleanup days" in some communities. Wastes are transported from collection stations to a network of 21 transfer stations. These stations also serve as collection points for industrial hazardous wastes that are not treated on-site by industry or sold for recycling. Transfer stations have a standard design and administrative procedure to ensure strict control of wastes, thereby avoiding transfer to sloppier and cheaper facilities. From the transfer stations, wastes are sent to the central Kommunekemi waste disposal facility at Nyborg (described in Chapter 3) for final treatment, destruction, or disposal.[10]

Although not formally required by law, most Danes make use of the collection stations because of their convenience. The stations accept free of charge paint residues, acids, spent solvents, and pesticides. They are conveniently located near existing community facilities such as fire stations or sewage treatment plants. Simply designed, they tend to consist of a small shed, drums for different types of waste, and a desk for keeping records of waste shipments.

Costs for the system are divided between the federal government, local government, and industry. The federal government takes an active role in financing the transportation and handling system because of its overriding interest in safe waste management. Local governments bear some of the costs for the collection stations. Costs for transfer stations are allocated to municipalities based on a formula that takes into account the origin of the waste, its quantity, the population of the community, and its degree of industrialization.

Kommunekemi, Denmark's central treatment facility, is jointly owned by private industry and various government entities. Construction was financed in part by an interest-free loan. Total collection and transportation costs in Denmark average about $60 per ton of waste in 1981 dollars. This relatively low cost, which has not changed significantly since 1981, is due to government subsidy. The subsidy is justified by the Danes as cheaper in the long run than the price of cleaning up abandoned hazardous waste sites.

In the West German state of Bavaria, seven collection sites accept waste from industry before treatment in three management facilities. The state government took the lead in providing for safe waste management by negotiating treatment and collection rules with both industry and citizen groups prior to implementation; and it owns a large percentage of the treatment facilities. Bavarian waste treatment facilities also feature relatively low disposal costs because of government subsidization of capital costs which, along with stiff penalties for illegal disposal, encourages industry to use the facilities. Householders are allowed to bring their toxic wastes to collection sites. Unlike Denmark, however, household waste collection is not promoted. This results from Bavaria's reliance on, and confidence in, household waste incineration.[11]

In Finland a government-industry partnership has constructed a central hazardous waste treatment plant. Modeled after Kommunekemi, the plant will be jointly owned in thirds by the state, municipalities, and private industry. It is planned as a non-profit facility, with fees set to cover treatment and residue disposal costs. Municipalities are given limited responsibility for collecting only oil wastes. Finland by-passes some of the intrinsic difficulties with packaged waste by leaving the responsibility for packing, storage, and transport to the producer of the waste. Industry also has the option of treating waste at the site of generation. Separate collection or transfer stations are not included in the Finnish system, nor are special provisions made for household wastes.[12]

An almost identical system is used in Sweden, where the responsible industry is required to transport waste to a central disposal plant. Waste oil, solvents, paints, acids, plating wastes, mercury wastes, cyanide, PCBs, and pesticides are all treated by separate processes. Again, private industry is not allowed in the waste disposal field on its own; disposal is a government-industry partnership. Sweden also has a network of transfer stations primarily for industrial waste. No special provisions, however, are made for household wastes at this time.[13]

Because of its lack of industrialization, large land area, and small population, Norway has steered away from a single large treatment plant and has concentrated instead on encouraging on-site treatment. So far, existing treatment capacity has proven to be adequate for the waste volumes generated. Nevertheless, municipalities are still obligated to maintain facilities that are partially subsidized by the Norwegian government for the collection of small quantities of hazardous waste. Wastes are containerized and stored until sufficient volume exists for shipment to a treatment facility. Norway is still contemplating the need for a single larger collection station.[14]

With the exception of Denmark and the West German state of Hessen, European countries have placed less emphasis on household toxic waste collection and disposal than on industrial hazardous waste processing. This may be due to the heavy reliance in Europe on solid waste incinerators and a belief that they provide the best possible control strategy for the low concentrations of toxics found in the residential waste stream. Most European countries will accept household toxic waste at collection centers, but do not particularly emphasize or promote this practice. Some Europeans are still not satisfied with municipal incineration as a safe solution for management of household waste containing toxic substances. This has been indicated by the recent moratorium on municipal incinerator siting in Sweden and in the West German state of Hessen and by collection efforts in Vienna, Austria, and Denmark.

A pilot program to collect and recycle both non-toxic wastes such as paper and glass and "problem wastes" such as motor oil, batteries, and medicines was conducted in Vienna in 1984. Organized by citizen volunteers with the help of city government, the program took place in two test areas, each consisting of about 10,000 people. Collection boxes for paper and glass were located on nearly every street corner. Each area neighborhood also had a "problem waste" collection center that would accept medicines, batteries, used motor oil, and pesticides. In addition, fifteen mobile "problem waste collection days" were held outside of the project area. On these days a wider variety of toxic materials, such as photochemical wastes, and caustic acids of diverse concentration, were collected. Program sponsors estimated that 22 percent of the population would use the problem waste collection stations if the initial results were projected over a full year. In contrast, only 6.3 percent of Vienna's households used one of the fifteen mobile collection schemes held. The sponsors estimated that total waste stream reduction from problem waste removal was less than 0.5 percent. The volume of problem wastes, however, was reduced by 30–60 percent. Some specific products such as batteries, potential contributors to mercury emissions at Vienna's garbage incinerators, were reduced by 50 percent. The project was deemed a success, and problem waste collection centers have been proposed for locations throughout the city. The key to success seems to be including problem waste collection as part of a larger waste collection and recycling system.[15]

European countries still, for the most part, concentrate on collecting and treating industrial hazardous wastes. Household substances are a secondary or tertiary concern. They are, however, successful in dealing with these small quantities of industrial wastes, due to government involvement and leadership. In the United States waste management decisions are made by industry directed and constrained only by economics and existing regulations. Moreover, the U.S. rule-making process tends to be long and contentious, often ending in litigation between citizen groups, regulators, and industry. This strategy has at times retarded the nation's efforts to move beyond dumping. By contrast, European rule-making begins as an arbitration process with various groups openly discussing management options and reforms before the promulgation of specific rules. This often encourages cooperation and agreement rather than polarization between parties,[16] as it lessens time spent on litigation and court delays. Interest-free loans, non-profit treatment facilities, and convenient collection stations have resulted in low costs in many parts of Europe; this encourages the use of proper facilities and minimizes the temptation to illegally dispose of waste. As explored at length in chapter 5, such cooperative ownership has the added advantage of making regulatory compliance easier to monitor and enforce because the government has immediate and ongoing access to industry records and operations.

ELEMENTS OF SUCCESS AND FAILURE

Nearly all state and local governments in the United States have chosen to leave the small generator problem to the EPA and Congress. States may have tightened the small generator exemption, but (with a few exceptions) they have not chosen to assist businesspeople with the management challenges of small waste quantities. Stemming in part from the free-enterprise tradition, America's small waste generators are still left to their own uncertain devices.

By contrast, many states and municipalities are attempting to deal aggressively with the small quantities of potentially toxic waste produced in the household. Household toxic waste is increasingly being viewed by agencies as simply one component of the larger municipal solid waste problem. As such, these wastes may be seen as requiring special handling; this usually occurs, for example, with yard wastes or bulky white goods (refrigerators and the like). Household toxics disposed of in the municipal waste stream are a public works problem and are therefore worthy of the attention of government programs. Householders are perceived as unlike the small commercial or business generators. Small generators are viewed as private entities responsible for their own waste disposal.

State governments in the United States have primarily confronted the problem of small quantities of household waste by sponsoring special days for the clean-up of household toxic waste. Small generators, however, are usually prohibited from involvement. Rules limiting the amount of waste that can be disposed of have kept businesses from using these services.

Problems in obtaining environmental liability insurance, siting difficulties, and neighborhood opposition to hazardous waste management sites are among the principal reasons state and local governments have failed to provide disposal services for small business generators. Regulations mandating strict technical requirements for collection sites have also discouraged collection programs serving the small business generator. Significant quantities of hazardous waste are thus excluded from collection in local assistance programs and may be disposed of inadequately instead.

Household collection days in Massachusetts are representative of programs held so far. A group of citizens becomes concerned over the number of potentially toxic substances in the household waste stream and is shocked by the lack of an alternative to disposing of toxics in the family garbage. The citizens then take the lead in organizing a household toxic waste collection day. A location and date is decided upon, publicity is generated, funding is obtained, and a hazardous waste hauler is hired.

Funding has proved problematic; local governments usually contribute only a small portion of the total amount. Programs are often dependent upon donations from business, industry, and local civic groups. Frequently an attempt is made to convince the hauler to operate the program at no cost as a public-relations gesture. Sponsors attempt to publicize the programs widely, brochures are distributed, and the media is involved as much as possible.

On the collection day itself, a desk is staffed by citizen volunteers who log in each waste item before it is given to the professional haulers. The waste is placed in appropriate drums by the hauler and taken to a treatment facility where each waste type is subjected to "appropriate treatment." Unfortunately, this often means that the waste is landfilled although occasionally incineration is used.

Household collection programs have generated good publicity for the sponsoring citizen groups, government, and industry, but so far they have removed an inconsequential portion of the waste stream. In Massachusetts, less than 2 percent of householders have participated in any given program; typical participation is even less than that.[17] Participation has improved somewhat in communities holding follow-up collection programs, but sponsors in Massachusetts noted:

The proportion of households participating was very low. The collection programs have not resolved the detrimental effects of these hazardous wastes on the environment. The majority of it is still being disposed of in landfills and sewers. However, the primary purpose of the collection program was one of education, and in that respect, organizers feel that they have been successful.[18]

The low participation rates experienced in Massachusetts are repeated in most communities holding collection days. In Seattle, for example, a pilot program attracted only 65 households; and in Hartford, Vermont, the 20,600–person community had only 109 people participate in a collection day.[19] Surveys distributed in Hartford indicated that 89 percent of the participants were college educated, substantiating other observations that participation is correlated with income and education. Low participation is seen by most sponsoring officials as a function of education and publicity, although most programs are routinely preceded by extensive publicity.

Lack of a convenient infrastructure is the reason for low participation in the United States. In Denmark collection stations are available each day in every community. Stringent regulations for permanent collection stations have nearly precluded their existence in the United States. RCRA requires that permanent stations be staffed, carry large insurance policies, and meet stringent technical standards. Only San Bernardino County in California has attempted to maintain permanent household toxic waste collection sites at several solid waste transfer stations. These sites have been operating under temporary RCRA Part A permits.[20] Current RCRA requirements for collection sites require funding and expertise beyond the capabilities of most local governments.

But lack of participation may be attributable to other factors as well. Even the best-publicized and most convenient collection program must still rely on what are perceived as individual acts of altruism. A more efficacious approach, given American political culture, might employ an economic incentive, perhaps in the form of a deposit on items such as pesticide containers or potentially toxic substances such as batteries. This would encourage broader-based citizen participation. A good example is provided by the "bottle bill" states. These areas experience in excess of 80 percent recycling of beverage containers because of the incentive provided by a simple cash-exchange return. This approach is already used in Maine, which requires a $5 or greater deposit on pesticide containers.

In addition to low participation, household toxic waste programs in the United States have proved to be quite expensive. In contrast to Denmark, where local government underwrites the cost of collection stations, funding in the United States usually comes from a hodgepodge of local, state, and federal grants, private and business contributions, and occasional volunteer work from waste disposal companies. Disposal costs for the Massachusetts programs average over $1,000 per ton; in San Diego and Redlands, California, costs were approximately $1,500 per ton; in Madison, Wisconsin, costs were over $8,000 per ton; in Irvine, California, $11,000 per ton; and in Anchorage, Alaska, due to high transportation costs, 87 individuals and businesses disposed of one-half ton of waste at a cost of $50,000.[21] Contrast these expenses with European facilities, where collection, transportation, and disposal costs in total rarely exceed $400 per ton.

The cost ineffectiveness of American collection programs can be traced to (1) the relatively high costs associated with setting up facilities for single-day collections; (2) transportation costs that usually involve hauling only a small amount of waste long distances to the disposal facility; and (3) low volumes collected relative to fixed disposal costs. Simply put, not enough waste is being collected to make the programs cost-effective in America. For a nation that prides itself on economies of scale, its blind eye to the collection benefits of an infrastructure is shocking as well as inconsistent.

In Denmark much lower costs are achieved by maintaining permanent collection stations at municipal facilities that are already staffed (for example, fire departments); by minimizing bureaucratic requirements (such as the $1 million liability insurance requirement typically found under U.S. rules, where the premium alone would break many small municipal budgets); and by holding wastes until a sufficient volume exists for economical transportation.

The lack of this infrastructure support is most dramatic in American states. Lynton Caldwell astutely summarized what is missing in America: "Effective policy is not merely a statement of things hoped for. It is a coherent, reasoned statement of goals and principles supported by evidence and formulated in language that enables those responsible for implementation to fulfill its interests." America simply still lacks this effective, coherent set of policies.

Lacking this infrastructure most American states subject communities to a myraid of requirements governing hazardous storage facilities if the collecting agents hold the wastes for more than 24 hours. Although this rule has served honorable purposes, it has hurt the development of collection schemes for small generators and virtually eliminated the prospects for household collections in most states. Thus the management of small quantities and household toxics remains in its infancy in the United States.

At least 23 separate pieces of legislation relating to toxic household products and chemicals are pending in 10 different states. This legislation includes such varying requirements as deposits on pesticide containers (currently mandated in Maine), the creation of study councils, funding for full-scale collection programs, and an excise tax on hazardous products.[22] Little has been done at the state level, however, to assist functionally and directly the small commercial and business generators.

Two states, Florida and Oregon, deserve particular mention because their approach to small-quantity waste management attempts to mirror the European example. In Florida the state funds a mobile waste collection station that travels from county to county collecting small quantities of household and business-generated waste on designated "Amnesty Days." In 1984 the first phase of the program collected over 250 tons of waste at 17 sites.[23] The program is, perhaps inadvertently, a mobile version of the Danish model. It is a milk-run service by which small generators are visited on a regular basis by waste transporters who collect and haul their wastes to a treatment center. Such a service has many obvious advantages to both industry and the general public since it creates a safe, convenient, and managerially efficient method of waste management, particularly for private industry.

The unknown factor in this Americanized milk-run concept is cost. Total collection, transportation, and disposal costs in the Florida program average $2,200 per ton. Again, we are reminded of the European example in which governments help subsidize collection systems because of the overriding social goal of avoiding environmental contamination. It is not surprising that the people in Florida, like the Danes, are very conscious of the fragility of their groundwater resources. It makes remarkable sense that Florida should be among the first states to design this progressive approach to the small generator problem. Other states, it strikes both authors, simply must follow this lead.

Oregon resembles Bavaria in that there are eight collection stations located throughout the state where small generators can bring their waste. Unlike Bavaria, however, the stations are privately operated by waste disposal companies who serve industrial customers and do not accept household toxic wastes. The Oregon Department of Environmental Quality recently proposed the construction of a number of small collection stations for household toxic wastes; the proposal has not been funded, and recent Oregon legislation required municipalities to conduct an assessment of how they could best manage their hazardous waste stream.[24]

THE NEED FOR UNIFORM ACTION

The need for assistance for the small generator continues to grow. The federal government's expansion of the universe of regulated wastes combined with the narrowing of the small generator exemption will create burdens for industry (the RCRA amendments alone double the law from 66 to 131 pages). If not properly anticipated, industries' difficulties in complying with the revised regulations may encourage improper high-risk disposal. State and federal governments should take the lead by providing the necessary advice and educational assistance (see Appendices A and B at the end of this chapter for private and public contacts).

The hazardous waste extension service previously described could serve as a model for state and federal action. In addition, state governments need to take a more activist role, similar to the European model, by funding and operating collection facilities. In the case of small quantity generators government has a responsibility for protecting public health and safety that can be met more effectively by stepping out of its role as adversarial regulator and moving into the role of a cooperating partner facilitating proper waste management.

In addition, state governments should take the lead in easing the liability restrictions on the collection of household toxics. Collection systems, such as those active in Denmark, could then prosper. Government already has an active role in managing solid waste; there is no reason household toxics should not be better managed along this public needs model of government leadership.

In the interim, an immediate opportunity to assist small business generators is the "milk-run" approach mentioned earlier. This concept may be best suited for rural areas where industries produce small waste quantities that can be stored until periodic pickup occurs. Many industries already do this, calling the hauler when sufficient waste volumes exist. Government should also take the lead in securing this service for small businesses by encouraging private haulers to provide it, by subsidizing the costs, if necessary, and as a last resort, by providing it directly.

Some recent developments show promise in America. The League of Women Voters (LWV), for instance, started a nationwide drive to institute a household hazardous waste pickup and disposal program on June 10, 1986. The LWV has circulated a videotape describing common problems encountered by community-based efforts. In addition, it has sent information kits to the media and various community leaders across the nation. Dow Chemical helped fund this program for the league.

Marcia Williams, director of EPA's Office of Solid Waste, has defined facilities dealing with household waste as "one of the highest priorities Congress has set" under the new RCRA amendments. In a recent statement published in *Pesticide and Toxic Chemical News,* Williams stated that regulations regarding household wastes are due by the spring.

Despite these recent advances, current small generator and household waste policies remain fragmented in the United States. They do not adequately respond to the intrinsic difficulties of small quantities management. A mixed set of resources has been directed to collecting household toxics, and little attention has been paid to small-business generators who collectively have large quantities of waste at significant toxic concentrations. Ignoring these markets simply aborts serious treatment opportunities. Several of the European countries noted above provide examples of comprehensive, effective systems for small-quantity waste management. The authors retain more detailed information for interested professionals. Where singular elements of these systems have been tried in the United States, they have met with mixed success. In short, the federal government needs to take a more active rather than reactive role in small-quantity hazardous waste management. To become prescriptive, U.S. government must accept a more cooperative role in solving waste management problems. For without a public works conception of itself, it cannot solve the future environmental catastrophies intrinsic to small-quantity and household toxics.

NOTES

1. S. Ridgeley and D. Galvin, "Summary Report of the Household Hazardous Waste Disposal Project Metro Toxicant Program 1," Metropolitan Seattle Service District, 1982.
2. Ibid.
3. A. A. Fungeroli, *Pollution of Subsurface Water by Sanitary Landfills, vol. 7,* U.S. EPA SW–12rg (Washington, D.C., 1971); U.S. Environmental Protection Agency, *Leachate Damage Assessment: Case Study of the Peoples Avenue Solid Waste Disposal Site in Rockford, Illinois,* U.S. EPA 530/SW–517 (Washington, D.C., 1976); U.S. Environmental Protection Agency, *Leachate Damage Assessment: Case Study of the Fox*

Valley Solid Waste Disposal Site in Aurora, Illinois, U.S. EPA 530/SW–514 (Washington, D.C., 1976).

4. C. F. Gurnham, B. A. Rose, H. R. Ritchie, W. T. Fetherston, and A. W. Smith, *Control of Heavy Metal Content of Municipal Wastewater Sludge,* National Science Foundation, Applied Science and Research Applications, NTIS PB–295–917 (Washington, D.C., 1979).

5. Ridgeley and Galvin, "Summary Report."

6. St. Lawrence County Environmental Management Council, *Draft EIS, St. Lawrence County Resource Recovery Facility* (Canton, New York, 1985).

7. Texas Department of Health, "The Small Quantity Generator of Municipal Hazardous Waste," Staff Paper, Bureau of Solid Waste Management, 1984.

8. B. Piasecki and J. Gravender, "The Missing Links: Restructuring Hazardous Waste Controls in America," *Technology Review,* October 1985, pp. 43–52.

9. New York State Hazardous Waste Treatment Facilities Task Force, *Final Report* (Albany, N.Y., 1985).

10. M. Palmark, "Collection Stations and Transportation Systems," paper presented at the First International Symposium on Operating European Centralized Hazardous (Chemical) Waste Management Facilities, Odense, Denmark, 1982.

11. CSB Herzogstv, "Disposal of Special Waste in Bavaria," GOB Munich 40, Bavaria, FRG, 1975.

12. F. Aarnio, "The Finnish System," paper presented at the First International Symposium on Operating European Centralized Hazardous (Chemical) Waste Management Facilities, Odense, Denmark, 1982.

13. L. Ljung "The Swedish System," paper presented at the First International Symposium on Operating European Centralized Hazardous (Chemical) Waste Management Facilities, Odense, Denmark, 1982.

14. Norway Ministry of the Environment, "A Study on the Nation-wide System for Collection, Transport and Treatment of Hazardous Waste," Report by the Steering Group, Oslo, Norway, 1981.

15. G. Vogel, *Erste Erkenntnisse aus der Alstion Planquadrat in Hinblick auf ein Integriertes Abfallkonzept für Wien,* government report, 1985.

16. Piasecki and Gravender, "Missing Links," pp. 49–52.

17. Massachusetts Bureau of Solid Waste Disposal, "Survey of Household Hazardous Waste Collection Programs," Boston, Massachusetts, 1984.

18. Ibid.

19. Vermont Agency of Environmental Protection, "Household Hazardous Waste Collection Day: Preliminary Assessment of Project," Montpelier, Vermont, 1984.

20. D. Duxbury, Tufts University Center for Environmental Education, personal communication with Piasecki.

21. Golden Empire Planning Center, *Household Hazardous Waste: Solving the Disposal Dilemma* (Sacramento, Calif., 1984).

22. J. Palenik and L. Rich, "A Legislative Look at Household Wastes," *Chemical Week,* August 21, 1985, p. 48.

23. Florida Department of Environmental Regulation, "Amnesty Days," Gainesville, Florida, 1984.

24. R. Wexler, Metro Portland Solid Waste Disposal, personal communication, 1985.

Appendix A
Household Hazardous Waste Collection Schemes: Critical Contacts in Existing State Programs with Select Corporate Sponsors

1. Alabama
   Gordon Kenna—Chemical Waste Mgmt., Inc.
   205–652–6721

2. Alaska
   Jim Sweeney—Alaska DEC
   907–564–1336

3. California
   Nancy Lancaster—California DHS
   916–324–0705

4. Colorado
   Orville Stoddard—Colorado DPH
   303–320–8333

5. Connecticut
   Leslie Lewis—Connecticut DEP
   202–566–3489

6. Florida
   Jan Kelman—Florida DEP
   904–488–0190

7. Indiana
   Danny Stubbs—GSX
   800–251–1227

8. Kentucky
   Keith Brock—Marion County
   502–692–3393

9. Louisiana
   Charles Goldsmith
   Dow Chemical
   504–389–6407

10. Maryland
    Mona Barteletti—GSX
    800–638–4440

11. Massachusetts
    Dana Duxbury—LWV and Tufts
    617–381–3486

12. Michigan
    Mike Keo—Dow Midland
    517–637–2646

13. Minnesota

Susan Ridgely—Minnesota P.C.A.
612–297–1453

14. New Hampshire
Chuck Knox—New Hampshire DPH
603–271–4622

15. New Jersey
Kevin Goshlin—New Jersey DEP
609–292–8341

16. New Mexico
Donna Lacombe
505–766–7434

17. New York
GECOS
716–873–4200

18. North Carolina
GSX Mark Johnson
800–334–5953

19. Ohio
CECOS-Pete Kmikilas
513–681–5738
Patricia Starr—LWV
216–499–3657

20. Oregon
Laurie Parker
503–229–5826

21. Pennsylvania
Karen Hoyt-Stewart—LWV
717–246–3486

22. Rhode Island
John Hartley—Rhode Island DEM
401–277–2797

23. Tennessee
GSX Danny Stubbs
800–251–1227

24. Vermont
Andy Rouleau—Vermont DEC
802–828–3395

25. Virginia
Fairfax County
Robin Bird
703–691–3381

26. Washington
David Galvin
206–447–5875

Dru Butler—Washington DOE
206–459–6308

27. Wisconsin
Leo Polczinski—Wisconsin DNR
608–266–5376

Appendix B
State Household Hazardous Waste Collection
Programs: Critical Political Contacts and Interested
Parties

1. California
Assemblywoman Sally Tanner
Aide—Dorothy Rice
916–445–0991

2. Colorado
Orville Stoddard—Colorado DPH
303–320–8333

3. Connecticut
Leslie Lewis—Connecticut DEC
203–566–3489

4. Delaware
John Sherman
302–736–5409

5. Florida
Jan Kelman—Florida DEP
904–488–0190

6. Maryland
Ann Sloan—Maryland HW
301–269–3291

7. Massachusetts
Anita Flanigan—Massachusetts DEM
617–727–3260

8. Minnesota
Susan Ridgely—Minnesota
612–297–1453

9. New Hampshire
Chuck Knox—New Hampshire DPH
603–271–4622

10. New Jersey
Kevin Goshlin—New Jersey DEP
609–292–8341

11. New York

Elizabeth Veanus—State legis.
518–455–3711

12. Oregon
Laurie Parker—Oregon DEQ
503–229–5826

13. Rhode Island
John Hartley—Rhode Island DEM
401–277–2797

14. Vermont
Andy Rouleau—Vermont DEC
802–828–3395

15. Washington
Dru Butler—Washington DOC
206–459–6308

16. Wisconsin
Leo Polczinski—Wisconsin DNR
608–266–5376

Concluding Remarks: The Next Steps after Land Disposal

Gary Davis and Bruce Piasecki

LESSONS FROM EUROPE

America now stands ready to dramatically redirect its hazardous waste management practices away from reliance upon land disposal. What lessons can be learned from Western Europe's fifteen years of experience with alternative technologies? Despite cultural and institutional differences, can some of the central ingredients of the European success story be used to design strategies appropriate to American political culture?

America began learning from the Europeans in the early 1980s. The Danish system and the Bavarian treatment facilities showed policymakers that there is no need to rely on direct land disposal, with its threat to ground water, since feasible and economic alternatives exist to recycle, detoxify, and destroy virtually all of the hazardous waste generated in this country. This lesson hardly seems novel today after the passage of the Hazardous and Solid Waste Amendments of 1984 which converted this attitudinal shift into law. Nonetheless, the United States still has a long way to go to use these alternatives comprehensively.

A further set of lessons remains to be summarized. These insights from Europe do not center upon superior technologies, but upon the political and institutional measures enacted by certain European governments that have virtually eliminated the use of land disposal for untreated hazardous waste. The chapters of this book present these essential lessons by describing European policies for waste reduction and waste management. Sifting through the many details of these policies discussed in the book, some key features emerge.

First, the European nations studied overcame the early preoccupation of hazardous waste management as a subset of garbage disposal by reconceiving the hazardous waste challenge as one of chemical engineering rather than one of dirt moving. This early recognition of the need to implement alternatives to land disposal probably stemmed from the fact that there is little vacant land for disposal of waste in Western Europe, and dependence on ground water for drinking water is widespread (up to 90 percent in some countries). Thus, obvious physical limits forced a shift in thinking more rapidly than in the United States.

The response of European policymakers to the hazarodous waste challenge focused on waste reduction and on securing the infrastructure of recycling and treatment facilities necessary to manage the waste by increasingly more stringent design and monitoring standards. This conception of the problem differs from the usual risk-management approach used for most other environmental problems, where government attempts to act as a neutral arbiter between public health and private business, proscribing harm as a regulator but rarely directing or carrying out solutions. America's repeated attempts to solve the hazardous waste problem in such a risk-management framework reveal some inherent features of the problem that do not lend themselves to this approach.

The hazardous waste problem is too intrinsically complicated to be solved exclusively by America's risk-management approach, since even the technical uncertainties are considerable. For instance, the difficulty in assessing the movement of contaminants from landfills has stymied risk managers. The focus on regulating landfill risks has also undervalued the issues of risk management for other technologies. The technical uncertainties, however, are minor compared to the institutional challenges. The difficulties, for example, of controlling packaged waste throughout its journey from generator to final resting place cannot be answered by risk assessment alone. At their core the decisions are political.

The European approaches described in this book center on political decisions to encourage waste reduction and to create and maintain an infrastructure of recycling and treatment facilities that use the best available technologies. This requires technical expertise as well as sustained political leadership. Although choices of technology and the risks of those techonologies are still important, the precise scientific justification for risk-based regulatory decisions may be less important than the organizational problems of creating and financing the infrastructure for improved hazardous waste management. Therefore, instead of quibbling over the risks of continuing to place specific types of hazardous waste in landfills, European nations such as Denmark have created hazardous waste management systems that provide incentives for waste reduction and that provide proven recycling and treatment technologies to industry as a public service.

A primary purpose of the Resource Conservation and Recovery Act, as the name of the act implies, was to facilitate waste reduction. This purpose was overlooked to the point that Congress very emphatically restated it in the 1984 amendments to the law. For over a decade waste reduction has assumed an increasingly central role in European hazardous waste policy, with European governments committing significant financial resources and technical expertise to solving pollution problems at their source. Waste reduction is referred to as "clean technologies" in Europe and programs include efforts to reduce all types of pollution (air, water, solid and hazardous waste) at the source in addition to conserving resources and energy.

As discussed in Chapter 2, the most important lesson concerning European waste reduction strategies is that they are perceived not only as environmental protection measures but also as integral parts of an industrial policy for revitalizing industry through increased efficiency and productivity. Given the small size of some of the countries studied, their level of support for "clean technologies" is remarkable and can be explained in terms of a conscious industrial policy to employ clean technologies as economic development tools.

The waste reduction orientation gives rise to dramatically different policy instruments from those of the traditional risk management regulatory approach. These include technical assistance, economic incentives to generators, technology transfer, and education programs. European waste reduction programs are not without a mandatory component, however, since several governments have the authority to mandate waste reduction in some fashion. Although rarely used, this authority provides strong incentives to insure the success of voluntary measures.

In addition to waste reduction, an original aim of the Resource Conservation and Recovery Act was to stimulate the growth of a private hazardous waste management industry by creating an attractive market of hazardous waste requiring licensed management. The market ended up being stalled, however, by the Environmental Protection Agency's dangerously sustained reliance upon land disposal. Now that EPA is charged with restricting the use of land disposal, it is finding that the infrastructure of recycling and treatment facilities may not yet be available to make the restrictions work.

European governments perceive hazardous waste management as a public need, much as highways and sewage treatment plants are viewed in America. As such, there is generally a high level of government involvement in the management of hazardous waste, ranging from the reported government ownership and operation of treatment facilities to the waste oil recycling subsidies discussed in Chapter 7. Government ownership or shared government–industry ownership in Europe has also been a reaction to a general distrust of private waste management firms, despite the fact that illegal dumping by private firms in Europe appears to be much less of a problem than it has been in the United States.

The shared government–industry ownership of treatment facilities discussed in Chapter 5 also displays a more cooperative relationship between government and industry in Europe. Industry has tolerated more government involvement at the same time governments have undertaken much of the responsibility and the liability for the management of hazardous waste. There is also more extensive informal consultation and consensus building with both industry and the public in the formulation of policy and regulations regarding hazardous waste and less resort to formal rulemaking or the courts to decide policy disputes, as noted in Chapter 9. This leads to less of a need for regulatory agencies to justify all of the details of a regulatory framework with detailed scientific proof of risks and cost/benefit analysis than has become common in the American system of environmental regulation.

American governments and businesses are now confronting the throny issues involved in the siting of hazardous waste treatment facilities that may be needed in the shift away from land disposal. While most European treatment facilities were sited before public concern about the risks of hazardous waste management had reached today's levels, much of the credit for the successful siting of these facilities can be attributed to the planning processes that preceded the siting attempts. Prior to singling out a site for these facilities, government plannners evaluated the need for the facilities and assessed the best available technologies that could be used. Because of government involvement in this planning process, these issues were dealt with in a public forum and a consensus generally existed about need and technology selection before a particulár site was selected. Although recent European siting attempts for large centralized treatment facilities have precipitated strong resistance despite the public planning process, government ownership has added flexibility to the siting process, resulting in more extensive public input concerning design and safety features that alleviated opposition.

European countries have recognized that the planning and siting of new facilities for the hazardous waste management infrastructure can sometimes conflict with the primary goal of waste reduction, since the availability of treatment facilities, and particularly those that are subsidized, provides a disincentive for hazardous waste generators to employ clean technologies. On the other hand, serious waste reduction diminishes the potential waste market for recycling and treatment facilities. Before overbuilding expensive hazardous waste treatment capacity in this country that may discourage waste reducation American policymakers should monitor European programs closely to see how those tensions are being resolved.

As a final noteworthy point, Scandanavian countries are pointing the way to an even more effective and efficient way of controlling the risks posed not just by hazardous waste, but by toxic chemicals in general. Integrated environmental management, as practiced in Norway, for instance, is a comprehensive evaluation of all toxic chemical discharges and emissions (to air, water, and land) from industrial sources. A single environmental permit that seeks to minimize overall emissions of toxic chemicals is issued by an environmental office that is not compartmentalized into medium-specific programs. Integrated environmental management prevents the solution of environmental problems in one environmental medium by shifting the pollutants to another medium, facilitating overall waste reduction and risk management. Integrated environmental management shows great promise and will be examined by the authors in future research.

AMERICAN ADAPTATIONS OF EUROPEAN MODELS

As several of the chapters state, it is unlikely that the European models of hazardous waste management discussed in this book will be adopted in identical fashion in America. Differences in culture, regulatory systems, and institutional frameworks are far too great. The European spirit of waste management is, however, growing in America and is manifesting itself in some uniquely American responses.

Technical Assistance Through a Waste Extension Service

The narrowing of the small generator exemption in the 1984 RCRA amendments means that over 795,000 more companies could be brought into the regulatory system, including many gas stations, dry cleaners, paint shops, and electroplating shops. These small firms will have difficulties in complying with the hazardous waste rules and in reducing their generation of hazardous waste because of a lack of expertise and resources to devote to waste management.

This predicament is similar to that confronting farmers early in this century. Agricultural techniques were clearly failing from overuse and inexpert soil management, but thousands of farmers lacked the information to upgrade their equipment and improve the handling of their land. Congress responded by passing the Smith-Lever Act of 1914, creating the Agricultural Extension Service. The act linked the federal Department of Agriculture with state land-grant universities and county governments and put county extension agents into the field to educate farmers about new techniques.

A similar example can be found in the energy field, where the oil crisis of the 1970s created the need to curb energy usage, requiring a reeducation of energy users and the application of some simple and effective conservation measures. Several electric utilities, such as the Tennessee Valley Authority and Pacific Gas & Electric, responded to this need by creating an energy audit service to perform free conservation audits for residential and industrial electricity customers providing relatively simple recommendations for conserving electricity, such as the use of insulation, weather stripping, solar hot-water heating, and cogeneration.

A hazardous waste extension service can similarly assist small- and medium-sized generators in implementing waste reduction measures, in proper handling of hazardous waste, and in compliance with the hazardous waste regulations. The extension service could serve as an arm of major universities in a manner similar to that of the Agricultural Extension Service, channeling the skills and research of faculty to the pressing needs of business. At least two such services are currently operating in universities in this country at the Georgia Institute of Technology and at the University of Tennessee. The Georgia Tech program, which has been funded by the U.S. Environmental Protection Agency for the last two years, made over 55 on-site visits during its first year of operation, focusing on improving compliance with hazardous waste regulations. The University of Tennessee program is in the first year of operation, and is intended to focus mainly on waste audits to suggest waste reduction measures.

Pollution Prevention Pays Programs: Expanding the Return on Investment Equation

The corporate slogan, Pollution Prevention Pays, coined by the 3M Company, has become a rallying point for a number of state programs to encourage waste reduction initiated over the last four years. These programs borrow heavily from the European spirit of cooperation and information sharing between government and industry.

The North Carolina PPP program began in 1983 shortly after the first pollution prevention conference to be held in the United States. The state program has since put together a compendium of pollution prevention success stories and has funded several research projects and symposia focusing on particular segments of industry important to the state.

By 1985 at least 15 states had some type of waste reduction program in operation or in the planning stages. These programs incorporate many of the measures discussed in Chapter 2 on European waste reduction programs. California, for instance, allocated $1 million for the first year of a program to provide grants to firms wishing to demonstrate innovative waste reduction technologies. Minnesota has created a Technical Assistance Program to assist waste generators in implementing waste reduction measures and also has a demonstration grant program. These and other state programs are expected to grow in the next few years, and to be assisted by the new national attention being given to waste reduction.

Waste Authorities and Treatment Utilities: The Promise of the Public Utilities Model

The European model of government ownership has not been embraced by the United States, despite a long and distinguished history of public works projects from dams to highways to sewage treatment plants. As pointed out in Chapter 5, the reconception of hazardous waste management as a public need in this country may require greater government involvement in creating a recycling and treatment infrastructure. Short of direct government ownership and operation of facilities, there are some American models that could accomplish the same purposes.

Although there have been a few government-owned hazardous waste landfill facilities operated in the United States, the closest model being implemented today is that of the waste authority. The waste authority is an independent, quasi-governmental corporation created for the explicit purpose of managing waste. It is imbued with some of the attributes of sovereignty possessed by governments in the United States, such as the right to condemn land or to issue tax-exempt bonds for financing. It is also operated as a nonprofit entity. The waste authority is actually very close to the European model, since many of the waste management companies discussed in Chapter 5 are independent nonprofit companies set up by the government.

One such waste authority exists today, the Gulf Coast Waste Disposal Authority, which operates in the Houston, Texas area for industrial and municipal wastewater treatment. For a number of years this authority has been considering the siting of an integrated treatment facility for hazardous waste generated in the area. Other waste authorities may eventually be utilized for construction and operation of treatment facilities in states such as Georgia, North Carolina, New York, Kentucky, and Minnesota, which have legislated the power to create such waste authorities but have not yet exercised this power.

Responding to the presence of a large private sector hazardous waste management industry in this country, most states and regional authorities have been reluctant to undertake ownership and operation of facilities. Nonetheless, Arizona and New York have undertaken processes to find suitable sites for, acquire the land for and then select private firms to construct and operate treatment facilities. New York's process stalled in its site selection phase, but Arizona has selected the site as well as the contractor and expects to have the facility under construction soon.

Can these waste authorities and government-sponsored treatment facilities compete in the private sector with other more profit-driven hazardous waste operators? This question is particularly important if these government-sponsored facilities employ more expensive treatment technologies before their competitors are required to do so by upgraded federal regulations. These facilities will not have the benefits of export restrictions or compulsory use that have been the linchpins in upgrading hazardous waste management in European nations using government-owned facilities.

An American institution that may provide answers to these problems is the regulated public utility, since it combines aspects of the private marketplace with the public works concept. Public utilities exist for electricity and gas service and for many other services. They operate as private for-profit corporations recognized as having an important public function. Because of this public function, a bargain is made between the corporation and the public. In exchange for providing reliable service in a specified manner at regulated prices, the corporation receives a guaranteed service area in which competition is restricted and a guaranteed return on its capital investment in the equipment necessary to provide the service.

A hazardous waste treatment utility would make the same significant bargain with the public. In exchange for treating all of the waste generated in an area, including hard-to-treat waste and household toxics, at set prices and using specified technologies, the private firm would have no competition in its service area and would receive a guaranteed return on its investment in treatment facilities.

This arrangement would have to be coupled with a statutory or regulatory requirement that generators in the service area deliver their hazardous waste to the treatment utility. Such a delivery requirement may be politically unacceptable, particularly where "cheaper" land disposal is allowed. The delivery requirement may also be challenged as violating the commerce clause of the U.S. Constitution, which prohibits states from erecting barriers to free trade across state boundaries. There is a precedent, however, in municipal garbage disposal "flow control," which means that municipal incineration facilities take automatic title to garbage within their service area and delivery is insured.

Transportable Treatment Units: Bringing the Solution to the Waste

A uniquely American answer to the questions of appropriate scale and the equities of siting large, integrated treatment facilities is rapidly emerging on the scene. Spurred by the need for permanent clean-up solutions at Superfund sites, several hazardous waste firms are developing and marketing transportable treatment units. The trailer-mounted units being offered include the full range of treatment technologies from rotary kiln incineration to sludge dewatering. In addition to their usefulness at Superfund sites, they will eventually be widely used in industry, traveling from one generator to another to treat hazardous waste as it is produced. This is already occurring for PCB oil in electrical transformers, which is chemically treated by a dechlorination process mounted on a truck trailer.

Portable treatment units are a solution to the siting conundrum described in Chapter 9, since they may alleviate the need for siting off-site commercial facilities to service those generators which cannot manage hazardous waste on-site. Several off-site treatment facilities have been stalled by communities that understandably object to becoming the center for treatment of hazardous waste from a wide area. Portable units would not create the same level of public concern as fixed commercial facilities, since they would only be in a community long enough to treat a relatively small quantity of waste that is generated exclusively by industries located in the community.

In addition to alleviating the siting problem, portable units offer the generator greater control over what happens to the waste, easing liability concerns since the treatment would be performed at the generator's facility. Finally, transporting the treatment unit to the waste would lessen the risks of transporting hazardous waste on the nation's highways.

Achieving a Common Agenda in Policy Development

While environmental rule making in the United States may never approach the level of cooperation common in many European nations, serious efforts are now being made at both the federal and state levels to deal with conflicts and disagreements before the formal part of the rule making process begins. Developing consensus in the early stages of the process may lessen the burden on the EPA and state agencies to justify decisions about risk with expensive research efforts and dubious cost/benefit calculations. However, Americans, from both the regulated community and citizen groups, will still insist upon due process in the development of regulations to ensure, on the one hand, that rule-making decisions are reasonably objective, and, on the other hand, that government agencies are not co-opted by the industries they regulate.

In the development of several of the regulations that are required by the 1984 RCRA amendments, EPA is using a consensus-building process on the front end before formally publishing draft regulations in the Federal Register. EPA also sponsored a process for developing a new regulatory approach to portable treatment units that would facilitate rather than hinder their use. This process involved two three-day meetings with a large group of representatives of citizen groups, hazardous waste generators, hazardous waste management firms, state regulatory agencies, and consultants. The group made recommendations to EPA for changes in the permitting process and in the Superfund remedial action selection process to facilitate the use of portable treatment units. EPA pledged to give the highest consideration to these recommendations in drafting new regulations for portable treatment units.

The states have also begun using consensus-building processes in development of controversial hazardous waste legislation. Virginia formed a toxics roundtable to develop its facility siting legislation. In a two-year process California utilized an official committee that included members of the legislature as well as other broad representation to develop hazardous waste planning legislation. Such processes are probably more common at the state level, where administrative processes are typically less formal.

Finally, there is some evidence that EPA is moving in the direction of the West German example of standard setting, outlined in Appendix 3, by providing the states with more input during the development of regulations that ultimately will be implemented by them. While EPA should not be constrained to wait for the states to agree on the need for and the substance of regulations to be applied consistently throughout the nation, the agency should probably formalize the input of state officials in a manner similar to the operation of the German Interstate Working Group on Waste (LAGA), especially since most of the practical experience in implementing hazardous waste regulations now resides at the state level.

Integrated Toxics Management: A Workplan for the Year 2000

Although the creation of EPA was an attempt to integrate environmental proctection in this country, that integration was only partial, since medium-specific regulatory programs have continued with different sets of regulations, different permits, and little coordination between regulatory staffs. This often-conflicting patchwork of regulations also has a few major holes, particularly where toxic chemical contamination is concerned. Yet modest steps have been taken toward integrated toxics management.

Most programs that strive for comprehensive management of toxics in this country are still in the information-gathering stage. The EPA Integrated Environmental Management Program has been conducting a series of case studies in different types of communities around the country to determine the extent of exposure of the public to toxic chemicals from various pathways. The goal is to determine if significant routes of exposure exist that are not being adequately controlled by current regulations and to determine what transfers of toxic pollutants are taking place from one environmental medium to another. State programs, such as New Jersey's Industrial Survey, attempt to determine the full extent of toxic chemical releases from industrial facilities into the air and water. Finally, Illinois is one of the few states to have a program specifically focused on nudging the state's regulatory programs toward integrated management of toxic chemicals.

The challenge for the next ten years of hazardous waste management in America is to once and for all break the long-standing dependence upon land disposal. The 1984 RCRA amendments have set the course, but their implementation will be difficult. First, the primacy of waste reduction can only be established with significant leadership and assistance from governments on both the federal and state levels. Second, the establishment of a safe and economical recycling and treatment infrastucture for waste that will be restricted from land disposal will require planning and involvement at all levels of government.

The most difficult task will be walking the fine line between the primary goal of waste reduction and the conflicting need to create and sustain the recycling and treatment infrastructure. It will also be difficult to avoid the appeal of perceived "quick fixes" such as ocean incineration and to proceed cautiously with land-based incineration while questions remain concerning the risks of incineration technology.

The Europeans have wrestled with these questions and have succeeded through significant government involvement in dramatically reducing the risks of hazardous waste management. Although many American companies are poised to profit from the new recycling and treatment market, careful attention to the European example will allow the United States to build upon the European successes in a manner appropriate to American culture and to avoid the failures.

Appendix 1

European Hazardous Waste Management Technology: Facilities and Personnel Narrative _____

GARY DAVIS AND BRUCE PIASECKI

INTRODUCTION

Piasecki and Davis spent the summers of 1983 and 1985 visiting European hazardous waste facilities and interviewing over 50 key regulatory officials, policymakers, and industrial representatives. Each interview, often scheduled with the assistance of the European Institute for Environmental Policy and the German Marshall Fund Staff, centered its concerns on the use of recycling, treatment, and destruction technologies for hazardous waste management. Frequent phone updating and correspondence has been transacted by both authors since, especially throughout 1986 and 1987.

Eight countries were visited in two three-month trips—Austria, Denmark, France, Holland, Norway, Sweden, Switzerland, and West Germany. In the course of these on-site inquiries, eight of the larger hazardous waste treatment facilities in five of these countries were investigated. Information was also collected about facilities not visited in each of the countries, as well as information about a new hazardous waste treatment facility in Finland. Key individuals with a broad knowledge of hazardous waste management in each of the visited countries were interviewed at length, and were often sent sets of follow-up inquiries through the mail. Updating of this information was further assisted by a grant to Bruce Piasecki by the New Jersey Hazardous Waste Facilities Siting Commission. The International Institute for Applied Systems Analysis, in Austria, also assisted the authors in developing their European contact list.

This appendix briefly describes the hazardous waste facilities in each of these countries. It also provides the names, addresses, and telephone numbers of useful professional staff

The authors wish to acknowledge the German Marshall Fund of the United States, as well as the Joyce Foundation, for financing the European travel and research resulting in this report.

at each facility or of individuals in the appropriate government offices who can arrange for meetings or correspondence with contacts at the most important facilities. The authors chose to assemble this appendix to illustrate the extent of the European experience in securing alternatives to land disposal.

Further information is available by contacting either Gary Davis at the University of Tennessee Waste Management Research and Education Institute, 327 South Stadium Hall, Knoxville, Tennessee, 37996, or Bruce Piasecki, Associate Director, Hazardous Waste and Toxic Substances Research and Management Center, Clarkson University, Potsdam, New York, 13676. For an account of European experiences in managing residues in above-ground land emplacements, contact Bruce Piasecki, President, American Hazard Control Group, 20 Hamilton Street, Potsdam, New York, 13676.

AUSTRIA

Austrian industries are estimated to generate approximately 300,000 tons per year of hazardous wastes. Although Austria has not had a comprehensive nationwide hazardous waste management program until recently, an incineration and treatment facility was built outside of Vienna in 1980 by a private firm called Entsorgungsbetriebe Simmering Ges.m.b.H. (EBS). Due to the lack of comprehensive controls over hazardous wastes, many waste generators shipped wastes out of the country, avoiding the higher-priced treatment at EBS in favor of cheap land disposal in other countries. As a result, EBS suffered financially and experienced operational problems. To stabilize the company and ensure the availability of hazardous waste treatment capacity for local industry, the city of Vienna took a controlling interest in the company in 1982.

EBS Facility, Vienna

The EBS facility consists of an incineration plant for hazardous wastes and waste oil, incinerators for sewage sludge from Vienna's sewage treatment plant, a physical/chemical treatment plant, and a wastewater treatment facility. The incineration plant consists of two rotary kilns, followed by afterburners, heat recovery boilers, and flue gas cleaning systems. It has a capacity of approximately 100,000 tons per year of organic solids (including wastes in drums), liquids, and sludges. The physical/chemical treatment plant was closed soon after operation began because of corrosion problems. A new facility is being constructed. The sewage sludge incinerators are two large fluidized bed combustors with a capacity of about 740 tons per day of sludge.

Prices at the EBS facility in 1985 ranged from about $25 per ton for incineration of used oil to about $250 per ton for incineration of pesticide wastes ($1.00 = AS 20). PCBs are not burned at the facility.

European Contacts

Dipl. Ing. Herbert Hofstetter
Manager
EBS
11. Haidequerstrasse
1110 Wien
Austria
Telephone: 43–222–76–16–10

Dr. Gerhard Vogel
Institut für Technologie und Warenwirtschaftslehre der Wirtschaftsuniversität Wien
Augasse 2–6
A-1090 Wien
Austria
Telephone: 43–222–34–05–25–801

Dr. Vogel is a university professor who has worked intimately with the Austrian government to design Austria's hazardous waste regulatory system. He is knowledgeable about all aspects of hazardous and solid waste management in Austria.

DENMARK

The Danish have probably the most comprehensive hazardous waste management system of any nation in the world. This Maryland-sized country generates approximately 100,000 tons of hazardous wastes per year. Most of these wastes are collected in 21 municipally owned waste transfer centers located throughout the country and transported to the Kommunekemi hazardous waste incineration and treatment facility near Nyborg. The plant is owned and operated by a consortium of Danish municipalities and has been operating since 1975. The total capital costs of the facility have been about $22 million, mostly provided by a low-interest, deferred-payment loan from the Danish municipalities. The Danish have been capitalizing on their experience in operating comprehensive hazardous waste management systems by consulting with other countries interested in developing management systems through a firm called ChemControl. The authors interviewed the principals of ChemControl on a number of occasions.

Kommunekemi, Nyborg

The Kommunekemi facility, which treats approximately 70,000 tons per year of hazardous wastes (including waste oil) consists of two rotary kiln incinerators and a physical/chemical treatment unit. A landfill for treatment residues is located on the coast a short distance away from the facility.

Incineration Plant

Kommunekemi currently operates two large rotary kiln incinerator systems, the first of which began operation in 1975. The second was started up at the beginning of 1982. The rotary kilns burn about 70,000 tons per year of solids, liquids, and semi-solids, including PCBs and other highly chlorinated organics. The rotary kilns are followed by afterburners, heat recovery boilers, and flue gas cleaning systems.

Capacity	approx. 90,000 tons/year
Kiln diameters	4.25 meters
Kiln lengths	12.0 meters
Operating temperature	1,200° C in kilns, 950° C in afterburners
Air pollution controls	Newer unit: novel wet/dry scrubbing system that contacts scrubber slurry with hot gases, resulting in no liquid discharges, also two electrostatic precipitators; older unit: one electrostatic precipitator

Air pollution standards met by the newest incinerator include the following:

HCl	300 mg/m³
HF	5 mg/m³
Sulfur dioxide	750 mg/m³
Dust	100 mg/m³
Pb	5 mg/m³
As	0.1 mg/m³

Actual emissions are significantly less than the standards.

Steam generated in the two waste heat boilers is used as process steam at the plant and to provide approximately 25 tons per hour (at 12 atmospheres) to the local district heating system for the city of Nyborg. Ash from the incinerators and dust from the flue gas cleaning system are disposed of at the nearby landfill.

Physical/Chemical Treatment Plant

The physical/chemical treatment plant at Kommunekemi treats approximately 7,000 tons per year of inorganic hazardous wastes, including acids, bases, and solutions containing cyanides and heavy metals. All treatment is performed in batch processes. Cyanides are destroyed by addition of sodium hypochlorite, hexavalent chromium is reduced to trivalent chromium by addition of iron sulfate, most heavy metals are precipitated by addition of calcium hydroxide, and mercury is precipitated by addition of sodium sulfide. The resulting sludges are filtered through a filter press, and the filtrate is pumped into a holding basin for analysis before release to the sewer system.

Discharge limits for water discharged to the sewage treatment plant include the following:

Cr (total)	2.0 mg/l
Hg	0.03 mg/l
Cd	0.05 mg/l
As	4.0 mg/l
F	10 mg/l
Nitrite	20 mg/l
CN	0.5 mg/l
Phenol	1.0 mg/l
COD	1,000 mg/1

(Chemical oxygen demand)

Toxicity   Guppie test—four *Lebistes reticulatus* should survive for 24 hours in filtered water diluted 15 times

The filter cakes are disposed of in the landfill and are segregated from other wastes so that recovery of metals can be accomplished if and when it becomes economical.

Waste Oil Plant

The waste oil plant at Kommunekemi treats about 20,000 tons per year of waste oil for the purpose of preparing it for combustion in the plant's incinerators. This facility should be of considerable interest to various American states, now suffering the lack of waste oil programs. The waste oil is heated to separate it into an emulsion phase and a sludge phase and to separate a volatile fraction. The emulsion phase is filtered to separate oil and water and the oil phase is further treated to separate volatiles. The oil, oil sludges, and volatile substances each are used to support combustion of other wastes in the incinerators. Wastewater from the process is used as a cooling agent in the afterburners of the incinerators.

Landfill

The landfill operated by Kommunekemi is approximately twelve miles from the plant near the sea and receives approximately 19,000 tons per year of wastes. The landfill is lined with a plastic membrane and has a leachate collection system. Groundwater in the area flows toward the sea and does not connect to any drinking water aquifers. There are two waste streams from the treatment plant—slag and dust from the incinerators and filter cakes from the physical/chemical treatment plant. In addition, a small amount of solid wastes is disposed of directly, such as oily earth or asbestos in containers. No liquids or toxic organics may be disposed of in the landfill. Filter cakes may not contain greater than 5 mg/kg of cyanide and 0.5 mg/kg of hexavalent chromium. Monitoring wells have shown no contamination from the landfill. For more information on this landfill, contact Bruce Piasecki

Prices

Sample prices for management of wastes at the Kommunekemi plant during 1985 are as follows ($1.00 = kr 10):

Used oil	Kommunekemi may pay for used oil
Solvents	$42–200/ton (charges are added for chlorine content)
Pesticide wastes	$320–390/ton
Inorganic wastes	$27–320/ton (depending on cyanide content)
Miscellanous	$8–450/ton (lower figure for oily soil; higher for highly chlorinated waste)

European Contacts

Peter Løvgren
ChemControl A/S
Dagmarhus
DK-1553 Copenhagen V
Denmark
Telephone: 45-1-141490

Mr. Løvgren is the chief spokesman for the Danish hazardous waste management system, and ChemControl provides consulting services to other countries and firms.

Per Rieman
Plant Manager
Kommunekemi A/S
Lindholmvej 3
DK-5800 Nyborg
Denmark
Telephone: 45-9-311244

FINLAND

Finland employed the Danish consulting firm ChemControl to design the Finnish hazardous waste management system along Danish lines. Construction was completed on a centralized incineration and treatment facility near the city of Riihimaki in early 1985 for the management of approximately 71,000 tons of hazardous wastes each year. The facility is owned by a non-profit company owned in equal shares by the national government, the municipalities, and industries generating hazardous wastes. The facility consists of an incineration plant, a physical/chemical treatment plant, a wastewater treatment plant, and a landfill.

Incineration Plant

The incineration plant is a single rotary kiln with afterburner, waste heat boiler, and gas cleaning equipment. Solids, liquids, and sludges may be burned, including highly chlorinated wastes. The bunker feed system, where solid and semi-solid organic wastes are mixed, has a nitrogen atmosphere to avoid explosive mixtures. The vapors are fed to the kiln.

Capacity   Approx. 35,000 tons/year
Air pollution controls   Novel wet/dry scrubbing system that contacts
                                 scrubber slurry with hot gases, resulting in no
                                 liquid discharges; followed by baghouse filters

The waste heat boiler produces steam for process uses and for the district heating system for the city of Riihimaki.
The following air emission standards must be met by the stack gas:

Dust	35 mg/m^3
HCl	35 mg/m^3
HF	5 mg/m^3

Physical/Chemical Treatment Plant

The physical/chemical treatment plant treats acidic and alkaline wastes, cyanides, chromates, and nitrates in batch processes. The capacity is 3,000 tons per year but can be tripled by running three shifts.

Water Treatment Plant

The water treatment plant treats oil/water emulsions, water containing organic compounds, and water containing low concentrations of heavy metals. Emulsions are treated chemically or with ultrafiltration, organic compounds are removed by activated carbon filtration, and heavy metals are removed by ion exchange. The capacity of the plant is 40,000 cubic meters per year.

Landfill

The landfill located at the facility is used for storage of treatment residues only. Its estimated capacity is 15,000 tons per year. For more information on Finland's residues strategy, contact Bruce Piasecki.

European Contacts:

Klaus Pfister
Ministry of Environment
Environmental Protection and Nature Conservation Department
Kaikutatu 3
P.O. Box 306
SF-00531 Helsinki 53
Finland
Telephone: 358–772–64–21

Matti Vattulainen
Oy Suomen Ongelmajäte (Finland's Problem Waste Management Company)
Telephone: 358–143–89–11

FRANCE

The French system of treatment and disposal facilities is privately owned and operated and dispersed throughout the country. According to a 1984 report there are 11 incineration facilities and 5 cement kilns burning hazardous wastes with a total capacity of 380,000 tons per year. There are 5 physical/chemical treatment facilities with a capacity of 240,000 tons per year and 10 facilities for the treatment of soluble oils and oil/water mixtures with a capacity of 340,000 tons per year. There are also 13 hazardous waste landfills and 18 solvent recycling facilities. Most of the facilities are heavily subsidized by the national government and by river basin financial agencies.

France currently has four integrated hazardous waste management centers, described briefly below.

TREDI Saint Vulbas Center

Saint Vulbas Center has been operated by TREDI since 1975. It contains three principal units.

Incineration Unit

The incineration unit is a drum-type furnace for incinerating solids with an afterburner and scrubber for cleaning combustion gases. Wastes burned include solid wastes such as contaminated packing material, chlorinated plastics, and paint wastes. The incineration unit has a capacity of 20,000 tons per year.

Physical/Chemical Treatment Unit

The physical/chemical treatment unit treats inorganic wastes, such as electroplating wastes, cyanide salts, acids and alkalines, and soluble oils (cutting fluids). Processes used include neutralization, precipitation, oxidation, reduction, and emulsion breaking. The unit also includes an ion-exchange resin regeneration system. The capacity of the unit is about 18,000 tons per year.

Sludge Dehydration Unit

The sludge dehydration unit has a capacity of approximately 5,000 tons per year.

European Contact

TREDI
Z.I. de la Plaine de l'Ain
St-Vulbas, 01150 Lagnieu
France
Telephone: 33–7–461–03–55

GEREP Mitry-Compans Center

The Mitry-Compans Center has been operated by GEREP-CA since 1977. It contains two principal units.

Incineration Unit

The incineration unit is specially designed to incinerate chlorinated or sulfonated liquid organics and solid wastes such as powdered plastic. The capacity of the incinerator is about 18,000 tons per year.

Physical/Chemical Treatment Unit

The physical/chemical treatment unit treats acid and alkaline wastes containing no cyanides or chromium. Its capacity is about 5,000 tons per year.

European Contact

GEREP
Z.I. de Mitry-Compans
77920 Mitry-Mory
France
Telephone: 33–6–427–16–97

SARP Limay Center

The Limay Center has been operated by SARP Industries since 1975. The total capital costs of the facility were about $12 million, approximately 40 percent of which came from a loan from the river basin agency. It receives approximately 140,000 tons per year of hazardous wastes, 90 percent from within 100 miles of the center. Costs at the center vary from about $18 to $180 per ton. It contains four principal units.

Incineration Unit

The incineration unit is a liquid injection furnace for the destruction of hydrocarbons, solvents, and other non-corrosive liquid organic wastes. Its capacity is about 15,000 tons per year.

Physical/Chemical Treatment Unit

The physical/chemical treatment unit treats cyanide wastes, acids and alkalies, heavy metal wastes, and laboratory wastes by neutralization, precipitation, chromium removal (using ferrous sulfate), and cyanide oxidation. It has a capacity of approximately 35,000 tons per year.

Sludge Stabilization Unit

The sludge stabilization unit solidifies treatment residues and other sludges using the CHEMFIX process. It has a capacity of about 100,000 cubic meters per year.

Copper Recovery

The copper recovery unit recovers copper in ammonia solutions from printed circuit etching in the electronics industry. Its capacity is about 3,000 tons per year.

European Contact

M. Gontard
President
S.A.R.P. Industries
Zone Portuaire de Limay
Porcheville—78250 Limay
France
Telephone: 33–3–092–04–77

TREDI Hombourg Center

The Hombourg Center, operated by TREDI, has been in operation since 1974. It contains four principal units.

Incineration Unit

The incineration unit is a small liquid injection furnace for destruction of liquid hydrocarbons, solvents, and pumpable organic sludges. It has a capacity of about 3,000 tons per year.

Physical/Chemical Treatment Unit

The physical/chemical treatment unit has the capability to neutralize and precipitate acidic and basic solutions containing heavy metals, to oxidize cyanides, and to reduce hexavalent chromium (by sulfur dioxide). Its capacity is about 28,000 tons per year.

Sludge Stabilization

The sludge stabilization unit solidifies sludges and has a capacity of about 30,000 tons per year.

Ion-Exchange Resin Regeneration

The unit is capable of regenerating about 1,000 cartridges per year.

European Contact

TREDI
Centre de Hombourg
BP 24
68490 Ottmarsheim
France
Telephone: 33–89–26–06–18

Other Useful Contacts in France

D. Lemarchand or Joseph Illand
ANRED
2 Square La Fayette
B.P. 406
49004 Angers Cedex
France
Telephone: 33–41–87–29–24

ANRED is the French national agency for the promotion of waste recycling, treatment, and disposal. The agency provides millions of dollars of grants and low-interest loans to waste management companies each year and provides technical and financial assistance to waste generators to help them reduce waste generation. Please see Chapter two for extended information on ANRED and its value for the U.S. policy development community. The two individuals listed here are knowledgeable about the French hazardous waste management industry and each of the hazardous waste facilities:

Yann Grenet
Ministère de l'Environnement
Direction de la Prévention des Pollutions
14 Boulevard du Général-Leclerc
92524 Neuilly-sur-Seine Cédex
France
Telephone: 33–1–75–81–212

The Ministry of the Environment is the national regulatory agency for hazardous wastes. Mr. Grenet is an engineer in the hazardous waste division who is knowledgeable about hazardous waste management in the country.

THE NETHERLANDS

Holland generates approximately 1 million tons per year of hazardous wastes. Over 50 percent of these are managed on-site, and of the quantity managed off-site, most is exported to treatment and disposal facilities in other countries, including Belgium, France, West Germany, and East Germany. There is one large hazardous waste incineration facility currently operating in Rotterdam, the AVR facility, which destroyed approximately 100,000 tons of hazardous wastes in 1985. AVR is a company jointly owned by the national government (10 percent), the city of Rotterdam (45 percent), and eight large industrial companies (45 percent).

The facility currently consists of one rotary kiln incinerator and one moving grid furnace. A new rotary kiln is under construction, which will have a capacity of 40,000 tons per year and will be able to burn highly chlorinated organics that are now burned at sea. Also of note, AVR-Chemie is developing concrete bunkers for the long-term storage of solid hazardous wastes, including treatment residues. (For designs and evaluations of the impacts of these bunkers, contact Bruce Piasecki.)

European Contacts

D. den Ouden
AVR-Chemie N.V.
Postbus 1120
3180 AZ Rosenburg
The Netherlands

J. Van Zijst
Ministry of Health and Environment
Division for Waste Management and Clean Technologies
Dkt. Reijersstraat 12
Postbox 450
2260 MB Leidschendam
The Netherlands
Telephone: 31–70–209367

NORWAY

Norway has taken a different approach to hazardous waste management than most other European countries. Since Norway generates relatively small quantities of hazardous wastes (about 120,000 tons per year) with several small generators spread out over a very large area, the government has opted not to construct centralized incineration facilities, but instead to rely upon existing industrial processes for hazardous waste destruction.

Most pumpable organic hazardous waste generated in the country is transported to a cement kiln for destruction. Wastes that cannot be pumped, but are not toxic or explosive, are destroyed in a boiler at a pulp and paper plant. Certain toxic organics that are not pumpable are mixed with water to form a slurry and then fed into the cement kiln.

Beginning in 1981, hazardous wastes were burned in a test program at a wet process cement kiln owned and operated by the Norcem cement company at Slemmestad, near Oslo. Due to declining demand for cement, the Slemmestad plant was shut down in late 1984, and the testing program was discontinued after demonstrating that hazardous wastes can be successfully destroyed in cement kilns with an efficiency greater than that of rotary kiln incinerators. Norcem is now constructing a full-scale hazardous waste feed system for an operating dry process rotary kiln located south of Oslo. The Norwegian government is financing the construction of collection stations for organic wastes to be transported to the kiln.

The test kiln at Slemmestad was 170 meters long with a diameter of 5 meters. In order to make cement, the temperature in the combustion chamber must be maintained at above 1400° C, and the residence time is measured in minutes rather than seconds because of the length of the kiln. Another advantage of using a cement kiln for burning hazardous wastes is that scrubbers are not needed to remove acid gases, since the alkaline cement clinker neutralizes the acids. Since cement production is a highly energy-intensive process, there is a strong incentive to replace some of the fuel normally used in a cement kiln (coal in this case) with hazardous wastes. Norcem also realized a profit from fees charged for the destruction of the wastes.

During the tests the kiln at Slemmestad was used to burn several different types of organic hazardous wastes, including PCBs and polynuclear aromatic hydrocarbons. Extensive air monitoring was performed for tests with these two hard-to-burn waste streams. For the PAH-containing waste stream, which was an oil refinery tar, the destruction efficiency was 99.99999 percent. Increased levels of aldehydes, alkanes, and alkylbenzenes were found in comparison to the burning of fuel alone, but all of these levels were very low. For the PCB test, the destruction efficiency was 99.99997 percent, and no chlorinated dioxins were found at detection levels of 0.5 ng/m^3. PCB emission concentrations were in the range from 0.08 to 0.17 μg/m^3, and the concentrations of organic compounds in the stack gas were increased a small amount. No PCBs were detected in the cement clinker or in the cement dust in the baghouse air cleaners. A similar test has been performed on the dry kiln that will be used for full-scale hazardous waste destruction. PCBs were not detected in the stack gas, but small amounts of dibenzofurans (1 μg/m^3) and dioxins (4 μg/m^3 total and 0.1 μg/m^3 2,3,7,8–TCDD) were detected.

European Contacts

Trygve Sverreson
Norcem
Haakon VII's gt. 2
Postboks 1386, Vika
Oslo 1
Norway
Telephone: 47–2–412770

Jan Johansen
State Pollution Control Authority
P.O. Box 8100 DEP., N-0032
Oslo 1
Norway
Telephone: 47–2–229810

Mr. Johansen is the head of the hazardous waste management division in the national environmental regulatory agency. He has close contact with the cement kiln program and developed Norway's hazardous waste plan, which stresses use of existing industrial processes for hazardous waste destruction.

SWEDEN

Sweden has followed the Danish model of hazardous waste management by establishing a government-owned company (SAKAB) responsible for managing most of the hazardous wastes in the country that are sent off-site for management. It has been estimated that approximately 480,000 tons per year of hazardous wastes are generated in Sweden and that over half of this total is managed on-site in facilities permitted by the environmental agency. To deal with a large portion of the wastes needing off-site treatment, SAKAB has established an incineration and treatment facility and a network of collection stations.

The SAKAB incineration and treatment facility, which began operating in 1984, is located at Norrtorp near the center of southern Sweden. The facility consists of an incineration plant, an oil recovery plant, a physical/chemical treatment plant, a mercury recovery plant, and a controlled landfill. The total capital investment for the facility has been approximately $46.5 million. It has an annual capacity of approximately 60,000 tons of hazardous wastes. The facility has gone through a rigorous testing and risk assessment phase with extensive monitoring during the licensing process.

Incineration Plant

The incineration plant consists of one rotary kiln for the destruction of liquid, solid, and semi-solid organic hazardous wastes, including PCBs. The system includes an afterburner, a waste heat boiler, and a flue gas cleaning system.

Capacity	Approx. 33,000 tons/year
Kiln diameter	4.5 meters
Kiln length	12.0 meters
Operating temperature	1,000–1,300° C in kilns, 1,000° C in afterburner
Air pollution controls	Novel wet/dry scrubbing system that contacts scrubber slurry with hot gases, resulting in no liquid discharges; followed by an electrostatic precipitator

Air pollution standards met by the incinerator include the following:

HCl	35 mg/m^3
HF	5 mg/m^3
Dust	35 mg/m^3

Physical/Chemical Treatment Plant

The physical/chemical treatment plant at SAKAB has a capacity of approximately 3,500 tons per year of inorganic acids and bases, and aqueous wastes containing cyanides and heavy metals. The plant uses standard processes for oxidation, reduction, precipitation, and filtration of these wastes. The filter cakes are deposited in a controlled landfill and the treated water is discharged into a surface water stream.

Oil Recovery Plant

Oily wastes are recovered based upon a process of heating to separate wastes into oil, water, and sludge phases. The oil is used as fuel in the incinerator and the sludge and water are burned. The capacity of the oil recovery plant is approximately 23,000 tons per year.

Discharge limits for water discharged to the surface stream from the physical/chemical treatment plant and the oil recovery plant are expressed in terms of kg/year and include the following:

Cr (total)	14 kg/year
Cd	5 kg/year

Cu	14 kg/year
Pb	14 kg/year
Ni	14 kg/year
Zn	14 kg/year
CN	0.4 kg/year
Oil	30 kg/year

Landfill

The landfill for treatment residues is located at the SAKAB facility. The landfill cells are built above ground upon a base of clay with a leachate collection system underneath. A metal cover is built over the cell to prevent rain from entering the cells. Finally, the groundwater table in the vicinity is lowered by pumping to prevent contamination. Groundwater is analyzed frequently. For a sustained evaluation of Sweden's above ground facility, see Piasecki and Ditz's report to the New Jersey Hazardous Waste Facilities Siting Commission on pretreated hazardous residues.

Prices

No price data was available when the facility was visited.

European Contacts

Lars Ljung
Technical Manager
SAKAB Norrtorp
692 00 Kumla
Sweden
Telephone: 46–19–77200

WEST GERMANY

West Germany has several integrated hazardous waste incineration and treatment facilities and relies heavily on treatment technologies for hazardous waste management. The technologies used in these facilities are proven technologies, and the West Germans have several years of experience in their operation. New technologies are being developed, particularly for recycling of metals and destruction of highly chlorinated organics. There is a growing concern in West Germany about toxic by-product air emissions (dioxins and furans) from the incineration of highly chlorinated organics, and few incinerators are permitted to burn PCBs. This concern is even greater when it comes to at-sea incineration. The Germans have virtually halted the at-sea incineration of hazardous wastes generated in their country.

A description of major hazardous waste treatment facilities and a list of contact people for each such facility follows.

ZVSMM, Schwabach

The Schwabach facility was probably the first off-site facility designed for hazardous waste disposal in the world. A clay-lined landfill with a leachate collection and treatment system was constructed for disposal of industrial wastes from the district of Middle Franconia in Bavaria in 1967 by a publicly owned corporation (ZVSMM). The facility was constructed at a time when most industrial wastes were disposed of with municipal wastes in unlined dumps and well before there were any hazardous waste statutes or regulations in West Germany. The facility operators recognized as early as 1967 that certain wastes should not be disposed of in the landfill and constructed an oil/water separation plant in 1968, a physical/chemical treatment plant in 1970, and a rotary kiln incinerator in 1970.

The facility today consists of one rotary kiln incinerator with afterburner and wet scrubbing system, the oil/water separation plant, the physical/chemical treatment plant, and the landfill. A new landfill has recently been opened at a site approximately 25 miles from Schwabach, and a second rotary kiln is planned for the facility. ZVSMM treats and disposes of approximately 120,000 tons of hazardous wastes per year. The total capital investment to date has been approximately $10 million. Operating costs for 1983 were approximately $3.2 million. The Schwabach plant was partially financed by the state of Bavaria and is owned and operated by the cities and counties of the district. (For a history of Schwabach's design and operations, contact Gary Davis or Bruce Piasecki.)

Oil/Water Separation Plant

The oil/water separation plant treats liquids and slurries containing oil with a maximum throughput of 48,000 tons per year. Approximately 35,000 tons per year are treated at the plant. The plant operates in three stages: (1) solid and phase separation with a mechanical decanter; (2) emulsion breaking by iron hydroxide; and (3) separation by sedimentation and filtration. A reactor for the chemical pretreatment of the oily wastes is also part of the system. The oil-bearing sludges are incinerated in the rotary kiln, and the aqueous solution, containing around 10 mg/l of oil, is discharged to the sewer system. ZVSMM has been experimenting with ultrafiltration units and planned to start using a full-scale unit in 1986.

Physical/Chemical Treatment Plant

The physical/chemical treatment plant treats inorganic wastes by neutralization, oxidation, reduction, and precipitation. Treatment is employed in batch systems. Cyanides are destroyed with hypochlorite, hexavalent chromium is reduced by sodium hydrosulfite, and basic wastes are often used to neutralize acidic wastes. The resulting sludges are settled and filtered with a filter press, and the filter cake is currently disposed of in the landfill. The aqueous stream from the treatment unit is only discharged to the local wastewater treatment plant after analysis, and reaction gases from the unit are scrubbed with a packed column using potassium permanganate. The maximum throughput is 36,000 tons per year. In 1984 approximately 10,000 tons were treated.

The water discharge permit for the discharge into the sewage treatment plant has the following limits:

Zn	3.0 mg/l
Ni	2.0 mg/l

Cu	1.0 mg/l
Cr (VI)	0.5 mg/l
Cr (total)	3.0 mg/l
Cd	1.0 mg/l
Pb	1.0 mg/l
Sn	3.0 mg/l
CN (free)	0.2 mg/l
Cl (free)	1.0 mg/l
Nitrate	30.0 mg/l
Phenol	5.0 mg/l
Hydrocarbons	20.0 mg/l

Incineration Plant

The incineration plant burns solid, semi-solid, and liquid organic hazardous wastes. The plant does not burn PCBs or other highly chlorinated organics because it has not been equipped with a scrubber and because the operating temperature in the kiln is limited by the refractory. A new wet scrubber will allow the unit to burn up to 10 percent chlorine. The scrubber blowdown liquid will be evaporated using waste heat from the incinerator, leaving a solid residue for disposal. Incineration ash is disposed of in the landfill.

Capacity	Approx. 18,000 metric tons/year
Kiln diameter	2.4 meters
Kiln length	8.5 meters
Operating temperature	900° C in kiln, 850° C in afterburner
Air pollution controls	Electrostatic precipitator for particulates, wet scrubber for acid gases

New air pollution standards for which the scrubber is being added allow the following:

HCl	50 mg/m^3
HF	2 mg/m^3
Dust	50 mg/m^3
Sulfur dioxide	200 mg/m^3
Nitrogen oxides	500 mg/m^3
Organics	20 mg/m^3
CO	100 mg/m^3
Hg, As, Cd	0.02 mg/m^3 in sum
Cr, Ni, Th	2.0 mg/m^3 in sum
V, Pb	
Cu, other metals	5.0 mg/m^3 in sum

Landfills

There are two landfills operated by ZVSMM, one located next to the treatment facilities at Schwabach and the other located at Raindorf, approximately 25 miles away. The Schwabach landfill is used only for treatment residues from the incineration and treatment plants. Approximately 70,000 tons/per year of these treatment residues are disposed of at the site. The Raindorf landfill will also accept treatment residues, such as metal sludges, from hazardous waste generators, contaminated soil, and other solid industrial wastes that cannot be disposed of with household garbage but that do not require pretreatment. Without pre-treatment, only non-toxic, neutral, and dewatered wastes may be disposed of.

The Schwabach landfill is approximately 61,000 square meters in area and can hold approximately 1.1 million cubic meters of wastes. The landfill is lined with at least 1 meter of clay and has a leachate collection system. Approximately 9,000 cubic meters of leachate are collected each year and treated in the physical/chemical treatment plant. There are eight monitoring wells, which are sampled four times per year. Recent monitoring has found a small increase in sulfates and chlorides in some of the wells, but the upgradient wells have shown the same increase.

The new Raindorf landfill is now being established after a lengthy siting process in which twenty sites in the district were investigated. The landfill will have a capacity of approximately 1.2 million cubic meters and will operate for around fifteen to twenty years. The landfill will also have a clay liner and leachate collection system. The leachate will be transported to the Schwabach facility for treatment.

Prices

Sample prices for hazardous waste management at ZVSMM facilities are as follows ($1.00 = DM 3.0):

Landfill	$30–65/ton
Oil/water separation	$23–46/ton (without pretreatment)
Physical/chemical	$17–180/ton (high figure is treatment for high concentrations of cyanides)
Incineration	$0–$200/ton (low figure for waste oils, high for chlorinated organics)

European Contacts

Hans-Georg Rückel, Director, or Dr. Norbert Amsoneit, Chief Chemist
Zweckverband Sondermüllplatz Mittelfranken
Postfach 1865
8540 Schwabach
Federal Republic of Germany
Telephone: 49–9122–7970

GSB, Ebenhausen

The Ebenhausen facility is the largest integrated off-site hazardous waste treatment facility in West Germany. It is owned and operated by GSB, a non-profit waste management company, to manage hazardous wastes in the major portion of Bavaria outside of the district of Middle Franconia. GSB is a partnership among the state of Bavaria, 75 industrial hazardous waste generators, and 3 local government associations. The company operates 10 regional collection stations, the Ebenhausen incineration and treatment facility, another incinerator for relatively non-hazardous industrial wastes, 1 landfill site, and 1 solvent recycling facility. In 1983 GSB managed approximately 228,000 tons of hazardous wastes generated by about 10,000 generators. Of this amount, approximately 150,000 tons were incinerated or treated at the Ebenhausen facility. Approximately 4,000 tons per year of solvents are recycled by GSB at a solvent recycling facility near Munich.

The Ebenhausen facility, constructed in 1976, consists of two rotary kiln incinerators, a physical/chemical treatment plant, and a wastewater treatment plant.

Physical/Chemical Treatment Plant

Physical/chemical treatment at Ebenhausen includes neutralization, precipitation, oxidation, reduction, and oil/water separation. Standard technologies are used. Cyanides up to 30 g/l can be oxidized by hypochlorite, hexavalent chromium is reduced, and the resulting sludges from the processes are dewatered with chamber filter presses and disposed of in the landfill at Gallenbach. Oil/water emulsions are treated by decanting, addition of emulsion-breaking chemicals, and vacuum drum filtration. The wastewater from the plant is held until analysis, then released directly to the Paar River or given further treatment in the wastewater treatment plant.

Discharge limits for water discharged to the river include the following:

Cl (free)	0.5 mg/l
F	50 mg/l
Nitrite	20 mg/l
CN	0.1 mg/l
Cr (VI)	0.5 mg/l
Cr (total)	2.0 mg/l
Hg	0.05 mg/l
Cd	0.5 mg/l
Pb	1.0 mg/l
Zn	3.0 mg/l
Ni	3.0 mg/l
Phenol	0.5 mg/l
TOC	50 mg/l

Incineration Plant

The two incinerators burn solid, liquid, and semi-solid hazardous wastes, and are comprised of two rotary kilns, a common afterburner with liquid injection burners, a heat recovery boiler, and air pollution control equipment. PCBs and highly chlorinated wastes can be burned at Ebenhausen, since the incinerator meets the standards of 1,200° C and three seconds residence time. The limit on chlorine content is around 10 percent. Electricity and steam generated by the waste heat supply all of the needs of the facility. Scrubber liquid blowdown is treated by the wastewater treatment plant and released to the river. Incineration ash is disposed of at the landfill.

Capacity	Approx. 65,000 metric tons/year
Kiln diameters	3.6 meters
Kiln lengths	12 meters
Operating temperature	1,200° C in kiln, 1,000° C in afterburner
Air pollution controls	Electrostatic precipitator for particulates, wet venturi scrubber for acid gases

Sampling for heavy metals in the fields around Ebenhausen has shown that high levels of mercury, cadmium, and lead are present. A new scrubber will be installed to improve removal of heavy metals.

Landfill

The GSB landfill that serves the Ebenhausen facility is located at Gallenbach, approximately twenty miles away. The landfill generally only takes treatment residues, dewatered sludges, and non-toxic solids. Small quantities of untreated wastes are allowed to be disposed of with permission from the state environmental agency. The landfill is located in a clay deposit, is lined with 40 centimeters of clay, and is built up in step fashion in a cut taken out of a hill. It is 6 meters above the first groundwater and 18 meters above a potable aquifer. Leachate is collected in two basins at the foot of the landfill and is treated before release into the river. Six groundwater monitoring wells are sampled periodically. The landfill capacity is approximately 1.4 million cubic meters and is expected to be used for another eighteen years. (Contact Bruce Piasecki for an extended assessment of Gallenbach and of its usefulness to American policymaking.)

Prices

Sample prices for hazardous waste management at Ebenhausen for 1985 are as follows ($1.00 = DM 3):

Landfill	$21–$65/ton
Physical/chemical treatment	$13–$80/ton (greater for high concentrations of cyanides or metals)
Incineration	$0–$206/ton (greater for PCBs and highly chlorinated organics)

European Contacts

Franz Defregger, Director
Waste Management Division
Bavarian State Ministry for Regional Development and Environmental Affairs
Rosenkavalierplatz 2
8000 Munich 81
Federal Republic of Germany
Telephone: 49–89–9214–2259

GSB
Herzogstrasse 60
8000 Munich 40
Federal Republic of Germany
Telephone: 49–89–3899–0

HIM, Biebesheim

The HIM hazardous waste incineration facility near Biebesheim in the state of Hessen is one of the world's newest and most sophisticated rotary kiln incineration facilities. The facility, opened in 1981, is owned and operated by a non-profit government/industry partnership. Wastes come predominantly from the state of Hessen, but are also imported from other West German states and from other countries. HIM also owns and operates two physical/chemical treatment plants in Kassel and Frankfurt and has planned a landfill near Mainflingen, which has been delayed due to public opposition. HIM manages a total of around 300,000 tons of hazardous wastes per year.

The incineration plant consists of two rotary kilns with separate afterburners, heat recovery boilers, and flue gas treatment. The plant burns solid, semi-solid, and liquid organic hazardous wastes, including PCBs. Incineration ash and solids from the flue gas treatment system are sent to the Herfa Neurode salt mines (see below) for long-term storage.

Capacity	Approx. 60,000 tons/year
Operating temperature	1,300° C in kiln, 1,200° C in afterburner
Air pollution controls	Novel wet/dry scrubbing system that contacts scrubber slurry with hot gases, resulting in no liquid discharges

Air pollution standards met by the incinerator include the following:

HCl	100 mg/m^3
HF	5 mg/m^3
Sulfur dioxide	200 mg/m^3
Dust	75 mg/m^3
Cd	0.14 mg/m^3
Pb	0.63 mg/m^3
Cr	0.24 mg/m^3
Cu	0.14 mg/m^3
V	0.0024 mg/m^3
Zn	0.96 mg/m^3

Steam generated in the two waste heat boilers is used to generate power and as process steam at the plant. Excess power is sold to the local public utility.

Prices

Sample prices for incineration at the Biebesheim incineration plant during 1985 are as follows ($1.00 = DM 3):

Used oil	HIM may pay for used oil
Solvents (low chlorine)	$133/ton (charges are added for chlorine content)
PCBs	$1,000/ton

European Contacts

Carl Otto Zubiller
Baudirektor
Hessischen Minesterium für Arbeit, Umwelt, und Soziales
Dostojewskistrasse 4
6200 Wiesbaden
Federal Republic of Germany
Telephone: 49–6121–817–2479

Hessische Industriemüll Gmbh (HIM)
Kranzplatz 11
6200 Wiesbaden
Federal Republic of Germany
Telephone: 49–6121–373–074

HIM-Biebesheim
Ausserhalb 34
6081 Biebesheim
Federal Republic of Germany
Telephone: 49–6258–6061

Herfa-Neurode Underground Disposal Facility

The Kali and Salz Company has operated an underground disposal facility for hazardous wastes in the mined-out portion of a salt deposit in the state of Hessen since 1972. The purpose of the facility is to totally remove some of the most toxic industrial wastes from the biosphere for geological time. The wastes are placed in mined "rooms" at a depth of about 700 meters. The rooms are sealed with brick, but the wastes can still be retrieved for recovery or treatment in the future.

Approximately 35,000 tons per year of wastes, such as cyanide salts, mercury sludges, chlorinated pesticides, and concentrated PCBs, are placed into the 300–meter-thick salt formation. No liquids or ignitable, explosive, or radioactive wastes are permitted. Most of the wastes come from West Germany, but a small portion are accepted from other countries.

The price for disposal of wastes in the Herfa-Neurode facility was approximately $60 per ton in 1983.

European Contacts

Dr. Gunnar Johnsson
Kali und Salz AG
Postfach 102029
Friedrich-Elbert-Strasse 160
3500 Kassel
Federal Republic of Germany
Telephone: 49–561–301–395

AGR Incineration Plant, Herten

The AGR Incineration Plant is located in the state of North Rhine–Westfalia, the West German state that generates the largest quantity of hazardous wastes (over 3.8 million tons per year). In contrast to the states of Bavaria and Hessen, the hazardous waste management system of North Rhine–Westfalia is mostly privately owned and operated. The AGR Incineration Plant, however, is owned and operated by a public authority in the heavily industrialized Ruhr River basin. The plant competes with private waste management firms that offer land disposal as an option. Thus the incineration plant has been operating at a fraction of its capacity, even with subsidized prices.

The AGR incinerator is a single rotary kiln with afterburner, heat recovery boiler, and flue gas scrubber, built in 1982 at a cost of approximately $30 million. The incinerator burns solids, liquids, and sludges, including PCBs and hospital wastes. The heat recovered from the incinerator is used to generate electricity for sale and steam for the local district heating system. AGR has been forced to build an evaporation system for the scrubber water from the incinerator, since a permit to discharge treated water into the Emscher

River was denied by the regional authorities. The evaporation system has experienced technical problems producing frequent down time for the plant.

The capacity of the plant is approximately 30,000 tons per year of hazardous wastes. The current, highly subsidized prices are about $60–80 per ton.

European Contacts

Dipl.-Geogr. Johann Fiolka
AGR
Ruttenscheider Strasse 66
4300 Essen 1
Federal Republic of Germany
Telephone: 49–201–7226–0

Other Useful Contacts

Dr.-Ing. Helmut Schnurer
Leader for the Working Group on Waste Management
Federal Ministry of the Interior
Graurheindorferstrasse 198, Postfach 170290
5300 Bonn 1
Federal Republic of Germany

Dr. Schnurer is the head of the federal hazardous waste management program for the federal government. His background is in nuclear engineering and nuclear waste management, so he has a good feel for the similarities between the nuclear and hazardous waste fields.

Dr. Ian Schmitt-Tegge
Umweltbundesamt
Bismarkplatz 1
1000 Berlin 33
Federal Republic of Germany

Dr. Schmitt-Tegge heads a group in the Federal Environmental Agency that deals with hazardous waste technology. The agency funds research into new technologies and acts as the technical arm of the federal government on environmental issues, although it has no regulatory authority.

Appendix 2

Obligations for Hazardous Waste Management in OECD Nations _____

BRUCE PIASECKI

Obligations by Country

I. OBLIGATIONS OF GENERATORS	1. Ban on Generation of Certain Wastes	2. General Obligation to Recover or Recycle Wastes	3. Waste Disposal Permits Required	4. Reporting Obligations	A. Initial Obligation	B. Annual Obligation	C. Regular Obligation	D. Obligation for Each Consignment	5. Obligation to Keep a Register	6A. Obligatory "Trip Ticket"	6B. Obligatory Consignment Note	7. Requirement for a Waste Manager	8. Obligation to Transfer Wastes to a Specified Site	9. Obligation to Supervise Transport and Disposal
International Community														
FRG			Y		Y	9	9	Y	Y	Y	28	Y	Y	31
Austria			Y		Y	9	9	9	Y	20	28		Y	31
Belgium			Y		Y	Y	Y	Y				Y		32
Canada						10	10	11		21				33
Denmark			Y		Y		Y				22		Y	34
U.S.			6		Y	Y		12	16	Y	28			33
Finland	1	4	6		Y				17			29		35
France	Y	Y	Y		Y		Y		Y	23				36
Ireland			Y						Y	Y	28			35
Italy		Y	Y		Y	Y			Y	Y	28			
Japan		5	Y		Y	Y			Y		24	Y		35
Luxembourg			Y					Y	Y	Y				37
Norway		4	Y		Y						25		Y	34
Netherlands	2	4	7				Y				26			
UK			8					Y	Y	Y				35
Sweden			6		13	Y	14		18			30	Y	
Switzerland	3	4						15	19	27	27			

Obligations by Country *Continued*

II. OBLIGATIONS OF CARRIERS OF HAZARDOUS WASTE	1. Obligation to Comply with ADR or Similar Rules	2A. Explicit Extension of Dangerous Goods System	2B. Packaging Obligations	2C. Specific Transport Equipment Obligations	3. Licensing of Carriers of Hazardous Waste	4A. Requirement of a Consignment Note	4B. Requirement for a Substance Identification Document	5. Obligation to Keep a Register	6. Obligation to Notify Authorities:	A. Initial Report	B. Annual Report	C. Regular Report	D. Report for Each Consignment	7A. Obligation to Provide Financial Sureties	7B. Obligation to Carry Special Insurance	8. Obligation to Handle Certain Wastes Separately	9. Explicit Obligation to Inform Authorities in Case of Accident
International Community																	
FRG	Y			38	Y	Y	62	Y		Y	54	54	54		Y		61
Austria	Y			38	Y	20	62	52		Y	54	54	54	Y			61
Belgium	Y		Y		44							55		Y	Y	61	Y
Canada		Y	Y	39	45	49	62	52		56			56	Y		61	Y
Denmark			40	40			Y										61
U.S.			Y	Y	46	Y	62	52		Y				Y		61	Y
Finland	Y	Y															61
France	Y					49	49	Y		57		58					61
Ireland						Y	62									Y	61
Italy	Y	Y			Y	Y	62	Y		Y	Y						61
Japan			41	41	Y		Y	Y		Y	Y						61
Luxembourg	Y		Y		47	Y	62	Y		Y				Y		Y	61
Norway	Y				48			53		59		59					Y
Netherlands	Y	Y					50										61
UK	Y		42	43		Y	62	Y						Y		61	61
Sweden	Y				Y		51								Y		61
Switzerland	Y						51							60			61

Obligations by Country *Continued*

III. OBLIGATIONS OF DISPOSERS	1. Licensing	2A. Obligatory Annual Reports	2B. Obligatory Regular Reports	2C. Obligatory Reports for Each Consignment	3. Obligation to Keep a Register	4. Record of Location for Each Deposit	5. Obligation to Appoint a Person in Charge of Surveillance	6. Special Closure Obligation	7. Post-Closure Surveillance	8A. Provide Financial Sureties	8B. Carry Special Insurance	9. Obligation to Treat Waste by Specified Methods	10. Obligation to Treat Certain Wastes
International Community													
FRG	Y	63	63	Y	Y		Y	71	Y	Y	75		Y
Austria	Y	68	Y		Y			71		Y	75		Y
Belgium	Y	68	Y		69		Y	72		Y	Y		Y
Canada	Y	64	64	Y	Y	Y	Y	73	Y	Y	Y		
Denmark	Y												
U.S.	Y	65	Y	65	Y	Y		Y	Y	Y	Y	Y	
Finland	Y				69		Y	72		76			
France	Y		Y		Y								Y
Ireland	Y			Y	Y	Y							
Italy	Y	Y			Y								
Japan	Y	Y			Y		Y	Y				Y	
Luxembourg	Y			Y	Y	Y							
Norway	Y	68	Y		Y	Y							Y
Netherlands	Y			Y	69			73		76			Y
UK	Y	63	63	Y	Y	Y							
Sweden	Y	66			69			74					Y
Switzerland	Y			67	70								

Notes for Obligations by Country Charts

1. The government is explicitly empowered to restrict the production, import, or use of a product that generates waste in the production process that cannot be made harmless and therefore causes excessive pollution.

2. Under section 34 of the Chemical Waste Act, the authorities have the power to enforce this obligation.

3. The Federal Act on Environmental Protection may be considered as implicitly prohibiting the generation of waste that could not be processed by facilities accessible to the generator of the waste.

4. The authorities are expressly empowered to institute recovery and recycling obligations.

5. Type of wastes are not defined.

6. Small generators, as defined in America, are still exempted.

7. Separate systems for generators and disposers are in place.

8. No permit is required for storage at this point.

9. This type of obligation may be imposed by specific regulations, or under the license to carry out the activities.

10. This type of obligation is usually imposed at the province level.

11. This type of obligation is laid down at the federal level for interprovincial and international transport. The provinces also apply consignment note systems requiring information to be supplied to the authorities.

12. Approximately half of the states had laid down this obligation by 1987.

13. This obligation results from the Environment Protection Act.

14. This obligation may be imposed pursuant to the Environment Protection Act.

15. This obligation is laid down in the draft ordinance on the subject.

16. However, there is a specific obligation to supply information on the waste disposal sites.

17. This obligation may be imposed through an administrative permission.

18. This obligation may be imposed pursuant to the Environment Protection Act.

19. This obligation is expressed in the current draft ordinance.

20. Required only for certain wastes.

21. At the provincial level there are obligations to complete manifests, and there is also a federal proposal for a standard document.

22. The waste must be accompanied by official identification forms.

23. There is, however, a trip ticket in draft form.

24. If a contract is concluded for waste processing or disposal, the generator must produce a document describing the nature and quantities of the waste concerned.

25. The waste must always be accompanied by a declaration to be completed by the generator on the composition of the waste.

26. The generator must fill in a form, have it completed by the disposal firm, then send it to the ministry prior to removal of the waste.

27. Under the current draft ordinance, a consignment note is to be drawn up by the supplier.

28. Superseded by the "Trip Ticket" requirements as of 1986.

29. The generator is obliged to do so only if he carries out the waste storage, processing, and disposal himself.

30. This obligation may be imposed pursuant to the Environment Protection Act.

31. Nevertheless, the generator must ensure that the waste is sent to a suitable disposal facility.

32. The generator is liable for any damage whatsoever caused by his waste.

33. Common-law fault liability may apply even following final waste disposal.

34. The concept of strict liability may be applied to hazardous waste generators before the courts.

35. The generator must ensure that the transporter is competent.

36. There is rather imprecise case law on liability of the generator regardless of any fault, thus existing precedents regarding dangerous goods in general must be taken into account.

37. The generator has obligations related to waste transport and handling, but these are passed on to the licensed collector as soon as the waste is handed over.

38. This obligation may be specified in the carrier's license.

39. This obligation is laid down in certain provincial regulations.

40. Similar obligations may be laid down by the Environment Protection Agency.

41. These regulations may require certain standards or rules to be complied with.

42. There are certain rules on the packaging of particularly hazardous substances for transport by road, and also general rules on the packaging and labeling of transported hazardous substances including waste that are due to come into force shortly.

43. There are rules on the marketing of vehicles and containers.

44. If the carrier is also the purchaser or importer of the toxic waste, a special license is needed.

45. This obligation is laid down at the provincial level.

46. A waste storage license is needed if the waste is to be held for more than ten days.

47. Collectors must be approved by stipulated official.

48. However, collectors of hazardous waste must obtain a permit.

49. Presently, declaration of loading in accordance with the rules for transport of dangerous goods; shortly (fourth quarter of 1984), hazardous waste consignment note according to article 8 of the Law of the 15th July 1975.

50. Obligations exist pursuant to the legislation on the carriage of dangerous goods.

51. Planned as of 1986.

52. Arises from "trip ticket" system.

53. Applies to waste collectors.

54. This obligation may be stipulated in the carrier's license.

55. However, purchasers and importers must be licensed to begin with and must submit reports monthly or even within eight days.

56. In most shipment situations carriers must be licensed at the provincial level. The carrier is also involved in the notification system by being required to complete a section of the manifest for each consignment.

57. A transport license is required, in accordance with the transport of dangerous goods regulations.

58. A quarterly report will be required as from the fourth quarter of 1984.

59. This obligation does exist, however, for the collectors of hazardous waste.

60. Some financial obligations may be applicable pursuant to legislation other than the Environment Protection Act.

61. Although not ruled explicitly, implicit in other regulations and practices.

62. Superseded by the "trip ticket" system as of 1986.

63. This obligation may exist in regard to the license or specific regulations.

64. This obligation may be imposed by the provinces. It is very likely that all hazardous waste disposal facilities will require a license to operate.

65. Half of the states already impose this obligation.

66. This obligation may be imposed pursuant to the Environment Protection Act.

67. The current draft ordinance imposes this obligation.

68. Superseded by requirements mandating only the Regular Reports system.

69. This obligation usually arises from the conditions in the license for carrying out disposal activities.

70. Planned as of 1986.

71. Waste disposers must report to the authorities when planning to close disposal facilities.

72. Some obligations still arise from the conditions in the license authorizing disposers to carry out their activities.

73. Some general obligations relating to the closure of facilities are laid down at the province level.

74. Obligations may be imposed under the Environment Protection Act.

75. Compulsory insurance is often a condition of the license.

76. Indirectly, disposers are obliged to have adequate financial resources.

Appendix 3

West Germany's Waste Catalog: Excerpts Including Preferred Management Methods _____

GARY DAVIS

Shortly after the passage of the West German Waste Law, an organization of state waste regulatory officials was created, called the Länderarbeitsgemeinschaft für Abfall (LAGA, Interstate Working Group on Waste). The LAGA compiled a waste catalog, which is a system of classifying all types of waste generated by households, industry, and commerce. The catalog contains 563 waste types, some classified according to chemical composition, but most classified according to origin. The LAGA catalog forms the basis for the regulation of waste management in the German states.

In 1977 the Federal Ministry of the Interior promulgated regulations listing 86 specific types of waste from the LAGA catalog as "special" or hazardous waste that could not be disposed of with normal household garbage. Many of the states, such as Hessen, with its three-tiered classification, have added several other waste types to this list.

Neither the Waste Law nor the original LAGA catalog said anything about how the waste streams were to be managed, leaving that supervision to be provided by the states in their waste plans. As a result, the states pursued different waste management strategies, ranging from the privately owned treatment and disposal facilities in North Rhine–Westfalia, emphasizing land disposal, to the state-owned centralized incineration and treatment facilities in Bavaria and Hessen described in appendix 1. These divergent strategies have created problems for states like Bavaria and Hessen that have invested millions of dollars in relatively expensive incineration and treatment facilities. Furthermore, as a result of

The authors wish to acknowledge the assistance of Carl Otto Zubiller, Director of Waste Management, Hessian Ministry of the Environment, Wiesbaden, FRG, in providing the information in this appendix and Mary Tannert of the German Department of the University of Tennessee for translation. For a history of Zubiller's role in establishing the waste catalog, write Bruce Piasecki, c/o American Hazard Control Group.

some states choosing not to pursue treatment methods for hazardous waste management, there is currently a shortage of incineration and treatment facilities in West Germany.

In 1982, in order to promote uniformity of waste management in the country, the Conference of State Environmental Ministers, at the urging of the state of Hessen and the city-state of Hamburg, directed the LAGA to develop uniform guidelines for the management of waste. The LAGA formed a special committee to perform this task, which included representation from the Ministry of the Interior and the Federal Environmental Agency. The committee has circulated its proposals extensively among industry, local governments, and citizen groups.

The first task of the committee was to formulate fundamental requirements specifying the minimum accepted environmentally sound technology for the management of each type of waste. The second task was to specify uniform standards for the design and operation of waste management facilities. The Federal Environmental Agency has been developing these standards based upon the best currently available technology with an attempt to build in flexibility to accomodate technical advances.

The uniform guidelines are still in the process of being finalized. When in final form, they will be promulgated as Technical Guidelines by the Ministry of the Interior and will be binding on the state governments. The following table contains the most important excerpts from the Waste Catalog for American purposes, including the draft preferred management technology designations specified by the LAGA committee. It should be noted that one other European nation, Austria, has also promulgated a waste list with preferred management technologies (Sonderabfallkatalog, June 1, 1983), given the force of law by the ordinance of February 9, 1984. With this Austrian attempt, West Germany's effort represents a major step forward for waste treatment. All the advanced industrialized nations of the world need take note of these two pathbreaking efforts to move beyond land disposal.

Excerpts from West German Waste Catalog with Preferred Management Designations[1]
(May 3, 1985)

Type	Origin	Preferred Management Method[2]
Rinse and wasnwater, organically satu- rated	Tank and container cleaning	1-P 2-I
Wire production residues	Wire production	1-I 2-D
Fatty acid residues	Production of edible fats, soaps	I
Oil, fat, and wax emulsions	Oil factories, production of soaps and cleaning supplies, wax goods, tank and container cleaning	1-P 2-I
Infectious manure	Institutes, production of pharma- ceutical products, cages of animals used for experimentation	1-I 2-P
Lime muds	Processing of rawhides	P or D
Tannery muds	Tannery, processing of rawhides	P or D
Sawdust and splinters oil-soaked	Soakup of solvents, accidents	I
Sawdust and splinters solvent-soaked	Soakup of solvents, accidents	I
Paper filters, oil- soaked	Oil purification, industrial vehicle repair shops	I

[1]The Waste Catalog is a listing of the whole range of solid and hazardous waste (special waste divided into two categories based on the relative hazard presented. This excerpt includes entries from category 2, the more hazardous category.

[2]I=Hazardous waste incinerator   S=Salt mine storage
 P=Physical/chemical treatment   M=Monofill
 D=Secure land disposal   L=Sanitary landfills

Type	Origin	Preferred Management Method
Paper filters, other-wise contaminated, mostly organically	Air and gas purification, filtration processes, chemical industry	I
Paper filters, other-wise contaminated, mostly inorganically	Air and gas purification, filtration processes, chemical industry	D
Packing material, contaminated or with dregs, mostly organic	Industrial activity	1-I 2-D
Packing material, contaminated or with dregs, mostly inorganic	Industrial activity	D
Lead dross	Lead foundries, printers	D
Light metal dross containing aluminum	Aluminum production, aluminum foundries, aluminum smelting plant	D
Light metal dross containing magnesium	Magnesium production, magnesium foundries, magnesium smelting plant	D
Zinc slag	Zinc production and foundries	D
Salt slag containing aluminum	Aluminum smelting works	1-S 2-D
Salt slag containing magnesium	Magnesium smelting works	1-S 2-D
Tin ashes	Tin production	D
Lead ashes	Lead production	D
Filterdusts, contain-ing non-iron metals	Non-iron metal production and steel, and malleable iron foundries	D

Type	Origin	Preferred Management Method
Electric furnace slag	Metal production	D
Blast furnace slag	Iron and steel production	M
Flue ashes and dusts from garbage incinerators	Garbage incinerators, sulfite waste incineration, sludge incinerators	1-M 2-D
Flue ashes and dusts from special waste incinerators	Special waste incinerators	1-D 2-S
Solid, salt-containing residues from flue-gas purification	Garbage incinerators	1-S 2-D or M
Solid, salt-containing residues from flue-gas purification	Special waste incinerators	1-S 2-D
Oil-contaminated soil	Oil accidents	1-D or P
Otherwise contaminated soil	Accidents	D
Used oil binders	Oil accidents	1-I 2-D or P
Used filter- and absorption-masses with production-specific admixtures (diatomaceous earth, activated carbon)	Chemical industry, chemical purification, adsorptive gas and fluid purification	I or D
Asbestos by-products asbestos dusts	Preparation and processing of asbestos, production of friction layers	D or L

Type	Origin	Preferred Management Method
Construction debris, earth excavation with production-specific admixtures, mostly organic	Building and facility demolition, oil and chemical accidents	D
Silicic acid and quartz by-products with production-specific admixtures, mostly organic	Industrial activity, ceramic industry, metallurgy, chemical industry	I
Silicic acid and quartz by-products with production-specific admixtures, mostly inorganic	Industrial activity, ceramic industry, metallurgy, chemical industry	D
Flue gas mud	Iron and steel production, iron, steel, and malleable iron foundries	D
Iron oxide muds from reduction	Chemical industry	D or M
Metal hardening room mud, containing cyanide	Metal hardening room	P or S
Metal hardening room mud, containing nitrates, nitrates	Metal hardening room	P or S
Barium sulfate mud containing mercury	Chemical industry, production of chlorine	D or S
Drilling muds with harmful contaminations	Deep drilling, drill sites, extraction of natural gas, oil and water	1-M 2-D
Calcium fluoride mud	Neutralization of hydrofluoric acid, emission purification, aluminum production	I-D 2-M or L

Type	Origin	Preferred Management Method
Iron metal packing and containers with harmful dregs	Industrial activity	I or D
Oil filters	Industrial vehicles, vehicle maintenance, machine shops	I
Lead by-products	Lead production, foundries printers, electrical technology, production of batteries and cables, lead processing	D
Beryllium splinters/ chips	Beryllium processing, production of navigation instruments	1-S 2-D
Lead-containing dust	Lead production and foundries, printers, electrical technology, production of batteries and cables, lead processing	D
Aluminum-containing dust	Aluminum production, processing foundries, and recycling smelting plants	P or D
Beryllium-containing dust	Beryllium processing, production of navigation instruments	I-S 2-D
Nickel-cadmium batteries	Production of batteries, trade and use	1-S 2-D
Mercury batteries	Production of batteries, trade and use	S
Dry batteries (dry cell)	Production of batteries, trade and use	D
Mercury, mercury-containing by-products, mercury vapor lamps	Production, trade and use; metallurgy	1-S 2-D
By-products with elementary sulfur	Chemical industry, production of rayon and dyes, gas purification	D

Type	Origin	Preferred Management Method
Electroplating mud containing cyanide	Electroplating	1-P 2-D
Electroplating muds containing chromium (VI)	Electroplating	1-P 2-D
Electroplating muds containing chromium (III)	Electroplating	P or D
Electroplating muds containing copper	Electroplating	P or D
Electroplating muds containing zinc	Electroplating	P or D
Electroplating muds containing cadmium	Electroplating	P or D
Electroplating muds containing nickel	Electroplating	P or D
Electroplating muds containing cobalt	Electroplating	P or D
Electroplating muds containing precious metals	Electroplating	1-P 2-D
Electroplating muds containing lead-tin	Electroplating	1-P 2-D
Other electroplating muds	Electroplating	1-P 2-D
Chromium oxide (III)	Chemical industry, production of pigments	D
Other metal hydroxide muds	Industrial wastewater purification, industrial activity	P or D or M
Tanning	Tanneries, rawhide processing slaughtering houses	1-S 2-D

Type	Origin	Preferred Management Method
Sodium and potassium-phosphate by-products	Chemical industry, production of cleaning products and preservatives	1-S 2-D
Wood preservative salt by-products	Wood preservation	1-S 2-D
Leather chemicals, tanning substances	Tannery	1-S 2-D
Fertilizer residues	Trade, use	S or D
Ammonium chloride	Chemical industry, soldering shops	S
Salt bath by-products	Warming baths, salt melting for heat convection	S
Ammonium bi-fluoride	Surface refinement of metals	S
Arsenic calcium	Non-iron metal production	S
Burnishing salt by-products	Surface refinement, production of tools and screws	S
Sodium bromide	Production of photochemical materials	S
Iron chloride	Pickling room, etching shop, chemical industry	S
Iron sulfate	Pickling room, etching shop chemical industry, production of pigments	S
Lead sulfate	Non-iron metal production, glass industry	S or D
Lead salts	Chemical industry, metal production	S
Heavy metal sulfides	Chemical industry, production of non-iron metals	D
Hardening salts containing cyanide	Chemical industry, hardening rooms	S

Type	Origin	Preferred Management Method
Hardening salts containing nitrates, nitrates	Chemical industry, hardening rooms	S
Vanadium salts	Metal production	S
Arsenic compounds	Chemical industry, glass and ceramic industries, metal production	1-S 2-D
Other soluble salts	Chemical industry	S
Other insoluble salts	Chemical industry	D
Used ammoniacal copper etching solution	Chemical industry, chemical trade, electrical technology	P
Battery acids	Industrial vehicles, federal railway, scrap metal trade	P
Inorganic acids, acid mixtures, pickling (acid)	Surface treatment of metals pickling, etching shops, electroplating, chemical industry, laboratories, hospitals, pickling rooms	P
Halogenated organic acids	Chemical and pharmaceutical industries	I
Halogen-free organic acids and acid mixtures	Chemical and pharmaceutical industries	1-I 2-P
Caustics, caustic mixtures	Surface treatment of metals, pickling rooms, etching shops, galvanizing plants, chemical industry	P
Ammonia solution	Photocopying shops	P or I
Hypochlorite waste	Cellulose production, textile industries, bleaching shop	P

Type	Origin	Preferred Management Method
Fixing bath	Photochemical operations, photo labs, x-ray labs, printers,	1-I 2-P
Sulfite waste	Cellulose production	I
Tanning liquor	Tanneries	1-P 2-I
Concentrates and half-concentrates containing chromium (VI)	Surface refinement	P
Concentrates and half-concentrates containing cyanide	Surface refinement	P
Wash and rinse water containing cyanide	Surface refinement	P
Bleaching baths	Film development and copying	P
Concentrates and half-concentrates containing metal salts (ie, nitrate solutions, rust removal baths, burnishing baths)	Surface treatment and refinement	P
Water and rinse water containing metal salts	Surface treatment and refinement	P
Copper salt solutions	Printers, etching shops, chemical industry, metallurgy, surface treatment	P
Iron salt solutions	Printers, etching shops, chemical industry, metallurgy	P
Developing baths	Photochemical operations, photo labs, printers	1-I 2-P
Remainders and old supplies of herbicides and pesticides	Chemical industry, production of herbicides, pesticides, trade and use	I or S

Type	Origin	Preferred Management Method
Production by-products of body care products	Production of body care products	1-I 2-D
Old medicines	Wholesale trade, pharmacies, hospitals, doctors' offices	1-I 2-D
Production by-products of pharmaceutical products	Production of pharmaceutical products	1-I or S 2-D
Waste oil	Gas stations, vehicle workshops industrial activity	1-P 2-I
Contaminated fuels (gasolines)	Tank farms	I
Transformer oils, thermal oils, hydraulic oils--free of polychlorinated biphenyls or terphenyls	Transformers, power substations, chemical industry, industrial activity	I
Transformer oils, thermal oils, hydraulic oils containing polychlorinated biphenyls and terphenyls	Chemical industry, power substations, transformers, mining	I or S
Electrical apparatuses containing PCB (polychlorinated biphenyls)	Production and use of transformers and condensators	I or S
Other waste containing PCB	Maintenance and removal of materials containing PCB	I or S
Oil sludge ("Olgatsch")*	Petrochemistry, paraffin oxidation	I
Fatty acid residues	Chemical industry, production of candles	Ø

Type	Origin	Preferred Management Method
Mud containing phenol	Petrochemistry, gas works, coke ovens	I
Distillation residues from creosote production	Petrochemistry, coke ovens, gas works	I
Ethyl chloride	Chemical industry, industrial activity	I
Chlorobenzene, free of PCB	Chemical industry, industrial activity	I
Chloroform	Chemical industry, industrial activity	I
Methylene chloride	Chemical industry, textile industry, degreasing of surfaces, removal of painting materials, plastics processing	I
Tetrachloroethylene	Chemical industry, textile industry, degreasing of metal surfaces, removal of painting materials, plastics processing	I
Carbon tetrachloride	Chemical industry, textile industry, degreasing of metal surfaces, removal of painting materials, plastics processing	I
Trichloroethane	Chemical industry, textile industry, degreasing of metal surfaces, removal of painting materials, plastics processing	I
Trichloroethylene	Chemical industry, textile industry, degreasing of metal surfaces, removal of painting materials, plastics processing	I
Halogenated solvent mixtures	Petrochemistry, industrial activity	I

Type	Origin	Preferred Management Method
Solvent-water mixtures with halogenized solvents	Chemical industry, chemical purification	I
Acetone	Chemical industry, textile industry, production of painting materials, plastics processing	I
Benzene	Cleaning and degreasing of metal surfaces	I
Cyclohexanone	Chemical industry, textile industry, production of painting materials, plastics processing	I
Diethyl ether	Chemical industry, production of pharmaceutical and pyrotechnical products	I
Methanol and other fluid alcohols	Chemical industry, production of pharmaceutical products, textile industry, production of painting materials	I
Methyl ethyl ketone	Chemical industry, textile industry, production of painting materials, plastics processing	I
Methyl isobutyl ketone	Chemical industry, textile industry, production of painting materials, plastics processing	I
Pyridine	Chemical industry, textile industry, plastics processing	I
Toluene	Cleaning and degreasing of metal surfaces, petrochemistry, coke ovens, gas works, chemical industry	I
Aliphatic amines	Plastics processing, chemical industry	I

Type	Origin	Preferred Management Method
Solvent mixtures without organic halogenized components	Petrochemistry, industrial activity	I
Muds containing solvents, with halogenized organic components	Chemical indusatry, metalworking, degreasing of metal surfaces	I
Lacquer and paint mud	Paint shop, spray booth, exhaust cleaning, enamel stripping	I
Organic coloring agents (pigments and dyes)	Production of coloring agents	I
Glue by-products, not hardened by precipitation	Production, trade, processing	I
Organic chemical waste from laboratories	Institutes, industrial labs, schools, chemical industry, trade	I
Distillation residues with halogenated organic components	Chemical industry, redistillation	I
Distillation residues without halogenated organic components	Chemical industry, redistillation	I
Polychlorinated biphenyls and terphenyls (PCB, PCT)	Chemical industry, PCB- and PCT-users	I or S
Phenols	Chemical industry	I
Infectious wastes	Hospitals and clinics with at least the following departments: blood bank, surgery, dialysis station, obstetrics, gynecology, isolation ward, microbiology, pathology, virology	I

Appendix 4

American Siting Initiatives: Recent State Developments ⎯⎯⎯⎯⎯⎯⎯⎯⎯⎯

MARY ENGLISH AND GARY DAVIS

INTRODUCTION

Approximately half of the states in the United States have legislated processes to guide and encourage the siting of "off-site" hazardous waste facilities. In all states with siting processes, the answers to three key questions mark the differences in their approaches: (1) Who initiates a facility proposal?; (2) Who decides whether to approve the proposal?; (3) Who participates in that decision?

The first section of this appendix reviews the answers that states have given to these questions. It is meant to convey the range of siting processes in the United States and common elements in these processes. No up-to-date, comprehensive compilation of state siting laws is currently available and changes may have taken place since the sources used by this review were compiled.[1] In addition, extensive changes can be expected over the next few years as states respond to the 1986 Superfund Amendments and Reauthorization Act. That act includes a requirement that Superfund cleanup activities shall be withdrawn unless a state has, by October 1989, entered into a contract or cooperative agreement with the president providing assurance of adequate capacity (either within the state or under an interstate agreement or authority) to treat or dispose of all hazardous wastes generated within the state over the next twenty years. How this provision will be interpreted is not yet clear, but it is likely to provide impetus for planning and siting activities.

The second section discusses how New Jersey, a state with formidable toxic waste problems, is approaching its siting process. Although the New Jersey approach should

The authors acknowledge with appreciation the reviews of Susan B. Boyle, New Jersey Hazardous Waste Facilities Siting Commission, and Barry Mitchell, Ontario Waste Management Corporation.

not be taken prescriptively, it illustrates two important concepts, discussed briefly in the concluding section and more extensively in Chapter 8. First, it suggests a shift from siting seen as a privately initiated marketplace activity, where the main dialogue is between the developer and the community, to siting seen as a publicly initiated activity with the state in a central role. Second, it suggests that a well-designed siting process is necessary but not sufficient for successful hazardous waste management. Simply following a specified procedure for siting a facility will not guarantee its approval or its economic viability, nor will it ensure that the facility has improved the overall management of hazardous waste. Instead, the other parts of a hazardous waste management process—needs assessments, waste reduction, and appropriate technologies—must work in concert with siting and will affect both the political success of the siting process and the practical success of the total hazardous waste management program.

SITING PROCESSES IN THE UNITED STATES TODAY

Who Initiates a Facility Proposal? Planning vs. Market Approaches

States are arriving at three different answers to this question. In most states the initiatory role has been and continues to be restricted to the developer, but in a growing number it includes the state, and in a few, local governments are encouraged to take an active part.

Market Approaches

All states with hazardous waste facility siting processes allow for the developer-initiated proposal. Within this option, there is considerable variation in the level of government to which the developer first applies. In many states (for example, Connecticut, Indiana, Kansas, Maine, Massachusetts, Michigan, New Hampshire, New York, Ohio, Rhode Island) the developer applies to either a line state agency or to a special state board. But in a number of states (for example, Colorado, Florida, Illinois, Kentucky, Maryland, North Carolina) the developer applies to the municipality or county first, and in at least two states (Iowa, Wisconsin) the developer notifies the state and local governments simultaneously. As will be discussed, who the proposal goes to first may set the stage for how it is received.

State Planning Approaches

Increasingly, states are taking the lead in their facility-siting processes. While this may be done as a backup if private proposals do not succeed or fill state needs (for example, Colorado, Connecticut, North Carolina), in most instances it is done in anticipation of siting difficulties and capacity needs. The approaches taken by the states vary.

Arizona, Florida, and Kentucky, for example, have allowed for developer-initiated siting processes, and each has also provided for a state-owned facility. In Arizona no private facility has yet been sited, but a legislatively approved site has been acquired for a state-owned facility, to be operated by a private contractor. In Florida, a needs assessment is now being done to determine the size and type of facility needed. Florida would prefer to stay out of the hazardous waste business but reserves the possibility of a state-owned facility if no private developer shows interest.[2] Similarly, Kentucky has established a Regional Integrated Waste Treatment and Disposal Facility Board to oversee the siting of one model multipurpose regional facility, but the board has determined that at present there is not a sufficient market to make the facility economically viable. Kentucky has reason to be cautious about economic viability. In Maryland, the Maryland Environmental Service (MES), a quasi-public corporation, was authorized to develop hazardous waste facilities subject to state permitting and site certification requirements, but the first MES facility, a landfill located at Hawkins Point in Baltimore Harbor, closed because it could not compete with existing facilities charging lower fees.[3]

Several other states (Minnesota, New Jersey, New York) are taking a different planning approach. Rather than developing needed facilities themselves, they are identifying sites for privately operated facilities. Minnesota and New Jersey are designating areas where developers would be encouraged to locate facilities, and Minnesota plans to actually acquire the sites. In the early 1980s New York also attempted to designate and acquire sites, but it was stymied by litigation and political pressure from the prospective host communities, who objected to its closed process and poor site selection.

Local Planning Approaches

States are beginning to actively engage their local governments in the front end of the siting process rather than having them simply react to proposals from outside firms. For the most part this is being done indirectly, through data collection intended in part to heighten a locality's awareness of its waste stream (for example, Massachusetts), or informally, by suggestions to local officials to consider how hazardous waste facilities can help attract and retain industry (for example, Virginia). But it is also being done more directly and formally. Florida, for example, requires that all counties and regional planning councils prepare hazardous waste management assessments, with each county designating areas where storage facilities may be located and each regional planning council designating possible treatment sites.[4]

California recently enacted a statute delegating to county governments the responsibility to comprehensively plan for the management of hazardous wastes generated in their jurisdictions. The plans, which are state-funded, must include waste characterizations and must identify the potential for waste reduction and recycling, the need for new off-site facilities, criteria to site the facilities, and general areas that might meet the criteria. No off-site hazardous waste facility may be sited or expanded unless the host county government determines that the facility is consistent with the plan.[5]

Who Decides Whether a Facility Will Be Sited? Preemption, Overrides, and Local Vetoes

In all but a few states, the state has the official final say on whether a hazardous waste facility site will be approved. Unofficially, local governments and citizen groups retain a great deal of power to block a facility through political pressure and legal challenges if they are adamantly opposed to it.[6] Whether they will be inclined to raise objections may depend in part on whether the local government has had an opportunity to make a decision, even if that decision is subsequently overridden, or whether its power to deny a project through the traditional means of local land use controls and local permits has been preempted.[7] Two other important distinctions involve the decision-making process within state government: first, who makes the final decision on a proposed site; and second, whether that decision precedes, follows, or is done in conjunction with the permitting process.

Preemption of Local Authority

Local governments are sometimes called "creatures of the state." All of their powers are transferred to them and in theory may be withdrawn. Traditionally, land-use control has been one of those locally vested powers, but few states have held to this tradition in their hazardous waste facility siting processes. Because of the perceived need for new hazardous waste facilities and the prospect of local opposition to such facilities, most siting laws provide that hazardous waste facility proposals shall bypass the local government and go directly to the state, with the state preempting the local power to accept or reject the facility.[8]

A few states (Connecticut, Massachusetts, Rhode Island, Virginia, Wisconsin) have attempted to soften the blow of state preemption by ensuring that the community can negotiate on the terms under which a facility will be sited (but not on whether it will be sited). As discussed in more detail later, this has met with mixed reactions.

Override

Only a few states (for example, Florida, Illinois, Maryland, California) allow the local government to make the initial decision on a proposed hazardous waste facility with some override powers retained by the state, and in at least one state (Illinois) the local authority must base its decision solely on criteria detailed in the state siting law.

Local Veto

On the other end of the scale, the siting laws of a few states allow local governments to retain much of their authority. In Colorado and Tennessee local governments may reject a facility even if it has been approved by the state, but they cannot allow a facility to be sited unless the state has determined that it conforms with general criteria set forth in state regulations. In Pennsylvania a local control approach has been used, but only as a temporary measure—after the state's hazardous waste plan has been completed, the state can override a local decision to reject a facility.[9]

Several states have hybrid approaches that gear who has the final decision and how that decision is reached to the kind of facility being proposed. Nebraska, for example, provides that the local governing authority has the final say over a proposed hazardous waste facility if it involves disposal, but if it does not, the developer simply goes through a state permitting process. Kentucky has a similar approach, but with the local veto limited to facilities involving land disposal.

State Siting Boards and Decisions within State Government

Although in almost all states the decision whether to issue a permit for a hazardous waste facility is made by the state agency responsible for administering the hazardous waste regulatory program, states vary as to who makes the final siting decision. In a few it is the governor (for example, Florida, North Carolina) or an existing state agency (for example, Kentucky, Nebraska, New Hampshire), but in most states it is a board composed primarily of private citizens representing different statewide interests. These boards (also called commissions or authorities) may include state agency representatives and often include representatives of the proposed host community. Most of the boards are permanent, with the members appointed by the governor for statutorily determined terms, and they usually are constituted to address only hazardous waste facility siting issues. However, in a few states (for example, Michigan, New York) they are ad hoc—that is, they exist only for the duration of the proposal under review; and in a few (for example, Illinois, Maine, Pennsylvania, Virginia) their missions are broader—for example, their scopes may include radioactive waste, non-hazardous waste, or environmental issues.

In most states the agencies or boards carrying out the siting or permitting processes are distinct but work in tandem. The extent to which the two processes overlap then becomes an important strategic question: if a facility permit is recommended or granted before the siting process is well underway, the prospective host community may see the state as biasing the siting process in favor of the developer; but if a determination of compliance with state regulations has not been made prior to the siting process, the state may be seen as endorsing a still-questionable project. Most states have attempted to resolve this apparent catch-22 situation by giving immediate notice of a proposed project to all affected parties; by making early, preliminary determinations of the project's technical soundness; and by holding off on any permits until a final siting decision is reached.

Some states, however, have either reversed this approach or integrated the siting and permitting processes. A number of states (for example, Connecticut, Indiana, Iowa, Kansas, Maine, Ohio, Pennsylvania, Rhode Island) require that the permitting agency give a notice of intent to approve the project's permits before the siting process is begun; a few (for example, Illinois, Kentucky, Michigan, New York, North Carolina) take the opposite course and stipulate that permitting must await an affirmative siting decision. New York, Ohio, and New Hampshire are trying the integrated approach: New York by providing that the permitting agency and the ad hoc siting board will conduct a joint adjudicatory hearing on all permits required, including the siting permit; Ohio by providing that, while a project's permitting review is to precede its siting review, the final permit is to be issued jointly by the siting board and the Ohio Environmental Protection Agency; and New Hampshire by providing that a single agency, the Office of Waste Management, issues a single permit, called a Site Certificate.

Nature of Public Participation in the Siting Process

This question is a source of great controversy. In the past two decades public participation has come to rank with motherhood and apple pie, but it has darker themes: Is it an insidious form of cooption? If it does not succeed in coopting, will it simply encourage opposition by providing a forum for dissent? But even if it builds dissent rather than consensus, should it still be included to ensure that all opinions are heard (or, more pragmatically, to try to ensure that opposition does not crop up later, in court)? These issues must be addressed, directly or by default, in any siting process. In addition, decisions must be made on whether the interests of the general public or those of the most immediate stakeholders will be stressed.

Most states with siting processes have adopted a traditional, two-part approach to public participation: (1) statewide interests are represented on a siting board, which, as discussed earlier, is usually a permanent decision-making body concerned solely with hazardous waste facility siting issues, with representatives of a prospective host community as temporary voting members; and (2) forums for state and local views on proposed hazardous waste facilities are provided through statutorily required public hearings and public comment periods. A growing number of states have gone further, in providing for public participation either by strengthening these elements or by adding other elements to the process, such as negotiation of siting agreements between the host community and the developer and grants to communities for independent reviews of site suitability.

Negotiation Approaches

The attempts of a few states (Massachusetts, followed by Wisconsin, Rhode Island, Connecticut, and Virginia) to formally recognize local stakeholder interests by establishing a developer/community negotiation process and by interposing the state as a mediator in that process have been widely publicized. Informed by the conceptual work of a group at the Massachusetts Institute of Technology and by a growing body of experience in environmental dispute resolution,[10] the 1980 Massachusetts siting act combines state preemption of local authority with mandatory negotiations between the developer and the community to reach an agreement on the technical and financial conditions under which the community would accept the facility. If no agreement is reached, the issue is subject to binding arbitration.[11]

The jury is still out on this approach—the Wisconsin, Rhode Island, Connecticut, and Virginia laws are as yet relatively untested. However, the experience under the Massachusetts law has so far been negative.[12] By 1985 five attempts had been made to site hazardous waste facilities, but all had been blocked by intense local opposition. The Massachusetts legislature is now considering a bill to substantially amend its siting process by adding statewide planning as a function of the state siting council, stressing the importance of waste reduction measures, making statutorily explicit the standards against which the council is to judge a proposed facility, strengthening the bargaining position of prospective host communities, and assuring their access to facilities for environmental monitoring purposes.[13]

While developer/community negotiation is not being abandoned, states are beginning to anticipate resistance to siting by adding earlier, more broadly based public participation mechanisms in planning processes. These generally have two purposes: first, to gain public acceptance of the concept of hazardous waste facilities, and second, to instill public confidence in the siting agency or board before any siting attempts are made. Widespread participation in the initial phases of the state's hazardous waste management program is emphasized, together with public education as an essential part of the program.

In Virginia, for example, a siting law was passed only after an extensive dialogue by a statewide group of environmentalists and industrialists (called the Toxics Roundtable) had, with a legislative commission on solid waste, hammered out a proposed bill. The subsequent siting criteria and other regulations were promulgated only after ten public hearings had been held around the state. In Minnesota and in New Jersey, public participation has been made an integral part of both the planning and the ensuing site-designation processes. But no facilities have yet been sited in these three states, so the success of their approaches remains uncertain.

THE NEW JERSEY APPROACH: THE PROMISE OF PLANNING

Those who have driven south out of New York City on the New Jersey Turnpike know that New Jersey is highly industrialized; they may also be aware that New Jersey is the most densely populated state in the United States. And they may have heard about New

Jersey's hazardous waste problems—about the 1,000 plus known or suspected disposal sites requiring cleanup (more than 90 of which are on the Superfund list) and the continuing stream of hazardous waste requiring off-site disposal (approximately 500,000 tons in 1983 alone, produced by about 2,000 generators). But they may be less aware of New Jersey's recent efforts to prevent further environmental degradation from improper disposal of its hazardous wastes.

To ensure that adequate hazardous waste treatment and disposal sites will be available in the future, New Jersey in 1981 enacted its Major Hazardous Waste Facilities Siting Act.[14] This act provides for the planning, siting, and licensing of new commercial hazardous waste treatment, storage, and disposal facilities by designating activities to be carried out by three agencies: the Hazardous Waste Facilities Siting Commission, the Department of Environmental Protection, and the Hazardous Waste Advisory Council.[15]

The Hazardous Waste Facilities Siting Commission

The 1981 act established the Hazardous Waste Facilities Siting Commission as a permanent body composed of nine members appointed by the governor with the advice and consent of the Senate. By law, the commission has three industry representatives, three county and municipal government representatives, and three environmental or public interest group representatives. When it is considering a specific site designation, two representatives of the affected county and municipality are to be added as ad hoc voting members. The commission's primary responsibilities include (1) adopting a hazardous waste facilities plan, (2) designating sites for new facilities determined by the plan to be necessary, (3) making siting decisions on facilities proposed for non-designated sites, (4) adopting rules and regulations for exemptions to the siting act, and (5) preparing environmental and health impact statements on proposed hazardous waste facilities. Assisting the commission is an executive director and a professional staff.

Beyond guiding the siting of facilities to handle the expected waste stream, the commission is also directed to examine ways in which this waste stream should be managed and reduced. The commission's proposed management hierarchy, in order of preference, includes source reduction, recyling, recovery, treatment and incineration, and disposal. (It hopes eventually to limit disposal facilities to residues from treatment processes, bulk wastes from the cleanup of contaminated sites, and a small quantity of wastes, such as metal sludges, for which there is no viable alternative.) To get advice on its reduction and recycling policies, it has formed a task force made up of representatives of environmental organizations, the academic community, and industry.

The Department of Environmental Protection

Under the siting act, New Jersey's Department of Environmental Protection (DEP) has the following responsibilities: (1) to adopt siting criteria after conducting public hearings and consulting with the Advisory Council; (2) together with the attorney general, to review the competency of applicants for hazardous waste facility licenses; and (3) to evaluate environmental and health impact statements associated with hazardous waste facilities, approve facility engineering designs, issue facility operating licenses, and conduct weekly inspections.

The Hazardous Waste Advisory Council

The Hazardous Waste Advisory Council is a thirteen-member body appointed by the governor with the advice and consent of the Senate. By law, its membership represents a wide range of interests, including environmental or public interest organizations, community organizations, municipal and county governments, firefighters, industries using on-site and off-site hazardous waste facilities, hazardous waste transporters, and hazardous waste facility operators. The council works with the commission on the plan, the site-designation process, and a public information program. It also advises the DEP on the siting criteria and on applications for hazardous waste facility licenses. It may hold public meetings on any siting-related matter and may review any matter at the request of the DEP or the commission.

The Siting Criteria

The act requires that siting criteria be designed to prevent any significant adverse impact on the public health or environment, including degradation of surface or ground waters.[16] During the criteria-development process, the DEP and the council met jointly over an eighteen-month period, and public meetings were held throughout the state on preliminary criteria. After considering public comments, revised criteria were issued, two rounds of public hearings were held, and the revised criteria were fine-tuned and then adopted by the DEP in September 1983.

The adopted criteria are classified by level: level 1 criteria have been used in conjunction with the DEP's computer mapping system to identify portions of the state that are excluded from further consideration; level 2 criteria have been used to carry out a similar negative screening on a regional (substate) scale; levels 3 and 4 are being used to identify candidate sites and to determine if they meet detailed performance standards. The criteria distinguish between land emplacement or impoundment facilities and other types of facilities, such as recycling and treatment facilities.

The Hazardous Waste Facilities Plan

Together with the siting criteria, the Hazardous Waste Facilities Plan provides the basis for the commission's siting decisions. The plan includes the following:

1. An inventory of all hazardous waste facilities within the state (including their life expectancy), and an identification of all people engaged in hazardous waste collection, treatment, storage, or disposal within the state

2. An inventory and a projection of the sources, composition, and quantity of hazardous waste to be generated within the state over the next three years

3. A determination of the number and type of new major facilities needed to treat, store, or dispose of hazardous waste in New Jersey

4. An analysis of the ability of all existing facilities to meet current and proposed state and federal environmental, health, and safety standards, and of the performance of the facilities in meeting these standards

5. An analysis of transportation routes and transportation costs from hazardous waste generators to existing or available suitable sites for major hazardous waste facilities

6. Procedures to encourage co-disposal of solid and hazardous waste, source reduction, materials and energy recovery, and waste exchange and recycling.

A draft plan was prepared by a consultant under contract with the commission and was released in 1984. Following a two-month public comment period during which five public hearings were held around the state, the plan was revised and released as final by the commission in March 1985. The act requires that the plan be updated every three years or more often as necessary.

From a comparison of projected waste with current capacity, the planning process determined that there is a need for two types of facilities: one 80–acre land emplacement facility and one or more rotary kiln incinerators.[17] It further determined that these facilities are needed both now and in the future, as New Jersey strives to shift management practices away from land emplacement and toward the hierarchy of practices that the commission has established as preferred.

Site Designation

The act provides for two companion approaches to siting hazardous waste facilities. Under the first, the commission performs its own study and designates a site or sites for each facility type called for in the plan. The commission then accepts proposals from commercial operators wishing to establish facilities at the designated locations. Under the second, the commercial operator identifies a site and proposes it to the commission. The commission then reviews it to determine whether it meets the siting criteria and is consistent with the plan and with other state objectives.

Under both siting approaches, once the Commission has made a preliminary determination about the site's possible acceptability, the prospective host county and municipality are notified and the municipality is given a grant to do a Municipal Site Suitability Study (MSSS)—its own assessment of whether the site meets the siting criteria.[18] After completion of the MSSS, an adjudicatory hearing is conducted by an administrative law judge, with the municipality allowed standing as a party of interest having the right to present testimony and cross-examine witnesses. The burden of proof rests on the commission (with a designated site) or the developer (with an undesignated site) to show that the site is suitable—that is, the judge must find clear and convincing evidence that the location of the facility will not be substantially detrimental to public health, safety, and welfare. After the judge's recommendations, one representative of the county and one representative of the municipality must be appointed to the commission, and within 30 days the commission must vote to affirm, conditionally affirm, or reject the judge's recommendations. The decision is considered a final agency action and is subject only to judicial review. Following the siting decision, the DEP proceeds with the permitting process.

Public Participation

In addition to public meetings and hearings on the siting criteria and plan, a public information program has been undertaken about the state's hazardous waste problems, the need for new facilities, and the siting process and its public participation opportunities. This program includes meetings accompanied with slide shows (for example, in 1985 the commission traveled throughout the state to explain its upcoming site search effort), brochures, a bimonthly newsletter, newspaper advertisements, and public service announcements about New Jersey's hazardous waste management problem and various strategies in dealing with it.

Where New Jersey Is Now

For New Jersey's siting process, the state's history of hazardous waste horror stories is a double-edged sword. People there are familiar with hazardous waste and its potential problems, and it takes little to convince them that the government must take the lead in doing something about these problems. However, this history may also have the effect of making people more alarmed about the prospect of having a facility in their town and more resistant to any superimposed site-designation process, no matter how carefully planned and painstakingly carried out.

Whether New Jersey's hazardous waste history will help or hinder its siting process may soon become clearer. In February 1986 the commission announced eleven candidate sites: four for a land emplacement facility and seven for one or two incinerators. These sites were to be assessed in detail by applying the level 4 criteria to on-site testing and were then to be ranked separately according to the type of facility being considered. As of November 1986, two sites had been tested, but both had failed level 4 requirements regarding groundwater movement. Access to most of the remaining sites was blocked by their owners, but suits to deny access have been unsuccessful, and the commission is planning to continue with the level 4 testing of all the sites.

In addition to the litigation over the site selection, some less anticipated reactions have cropped up. (1) Although the statewide plan was well received initially, people in the candidate localities have questioned whether the facilities it calls for are truly needed. Not surprisingly, they feel that full-scale waste reduction efforts should be made before any facility is sited. (2) More surprisingly, people from one municipality have expressed the opinion that they would prefer to negotiate with the developer, not the commission.[19] But both the developers and the municipalities are hedging their bets: the developers are awaiting the outcome of the site-designation process—only one has proposed a site—and until this outcome is known, the municipalities are likely to be cool toward any developer-initiated proposal. (3) The targeted municipalities are questioning whether they can count on the state to continue carrying out facility inspection and safety enforcement measures, rather than abandoning them once the facility is in place. As a result, some are pointing to the need for municipal personnel to be trained in inspection and monitoring techniques and are asking for the right to close a facility down.[20]

The outcomes of these and other obstacles to New Jersey's siting process cannot be foreseen. If the process hits snags, it certainly will not be for lack of effort: New Jersey has diligently tried to devise an approach that is planned rationally and carried out with extensive public involvement.

STATE LEADERSHIP IN SITING

The New Jersey approach exemplifies a growing trend in state siting efforts—a trend away from the siting process seen as a dance (or battle) between the developer and the community with the state in a mediating role, and toward the siting process seen as an activity in which the state plays a central role by determining what treatment, storage, and disposal facilities are needed, where they might be located, and how they would fit into a total hazardous waste management program. The proposed changes in the Massachusetts siting law also reflect this trend.

This approach to siting is closely akin to the European approaches discussed in Chapter 8, which have been sustained by a consensus about the need for centralized facilities. But, as discussed in that chapter, the European consensus appears to be breaking down. The difficulties that have arisen in New Jersey's well–thought-out process may be indicative of a similar problem. The most important equity issue with hazardous waste facilities—the issue of dispersed benefits and concentrated risks—may be the biggest block to siting a centralized facility.

By fostering planning and siting on a more local level, states such as California and Florida have attempted to address this equity issue. Rather than having hazardous waste management be wholly a state function, local governments are asked to assume part of the responsibility. It remains to be seen whether their conclusions about the types and sizes of facilities to be sited will meet aggregate statewide needs, and whether the ground-up process will avoid the equity objections that are being raised to the top-down process.

This country has begun to awaken to the fact that safe hazardous waste management is a public need, and as a result it is experiencing a trend away from siting seen simply as a marketplace activity. It is also experiencing a trend toward regarding siting as only one part of an overall hazardous waste management system, rather than as a singular, all-important exercise. But only time will tell whether either state or local efforts can successfully site the facilities that may be needed to provide viable alternatives to land disposal.

NOTES

1. Ann Sprightly Ryan, "Approaches to Hazardous Waste Facility Siting in the United States," unpublished paper available from the Massachusetts Hazardous Waste Facility Site Safety Council, September 1984; National Council of State Legislatures, Solid and Hazardous Waste Project, *Hazardous Waste Management: A Survey of State Legislation, 1982.* (Denver: National Council of State Legislatures, 1982) (An update of this report is to be released in 1987). National Governors' Association, "Abstracts of State Hazardous Waste Siting Laws," unpublished material available from the National Governors' Association, February 1982.

2. According to Richard Deadman, Florida's siting coordinator, the state's isolated location and the diversity of its hazardous wastes may detract from the economic attractiveness of a hazardous waste facility (personal communication, September 26, 1986). Other states may encounter similar problems: for example, Iowa's Select Advisory Panel on Hazardous wastes has commented that "the quantity of hazardous waste produced in Iowa is not sufficient to attract private development of a facility. If a management facility is needed, the State will have to help fund it." Iowa Department of Water, Air and Waste Management, *Hazardous Waste Management Plan*, February 1985, VI–2.

3. Richard N. L. Andrews and Terrence K. Pierson. "Local Control or State Override: Experiences and Lessons to Date," *Policy Studies Journal* 14, no. 1 (September 1985): 94.

4. Richard Deadman has indicated that some counties and regions gave this require-

ment only a pro forma response by designating patently unsuitable sites, but that the requirement did increase public awareness of locally generated waste streams (personal communication, September 26, 1986).

5. California Assembly Bill no. 2948, signed October 1, 1986.

6. While these court challenges are not always successful, a developer may need great perseverance to see them through. For example, an April 1984 decision by the Ohio Hazardous Waste Facility Board to issue a permit to Waste Technologies Industries for a storage and treatment facility in East Liverpool was appealed to the Franklin County Court of Appeals. In December 1985 that court upheld the board's decision, but the case was then taken to the Ohio Supreme Court, which has agreed to hear the appeal.

7. This point was suggested by David Morell's comment in "Siting and the Politics of Equity," *Hazardous Waste*, no. 4 (1984): 561, that "a siting process based on state override seems to have a much greater chance to gain local respect than does a process based on state preemption. This distinction is admittedly slight, and rather subtle. The fundamental principle involves the explicit grant of authority by the state legislature to the local community to say 'yes' or 'no' to the proposed hazardous waste facility. A local decision of this kind is vastly different from local participation on state preemptive decisions. . . . Citizens feel isolated from such decisions. They resent the way in which preemption places priority on administrative efficiency over political sensitivity. They become angry at the evident lack of state accountability to local interests."

8. The state may also use preemption to bring flexibility to siting regulations. In Virginia, for example, the Hazardous Waste Facility Siting Council (now the Virginia Waste Management Board) may use its discretionary power to reduce the siting requirements and fees for small-scale storage facilities (Harry E. Gregori, Jr., Executive Director of the Siting Council, personal communication, March 21, 1985).

9. On July 15, 1986, Pennsylvania's Environmental Quality Board adopted a Hazardous Waste Facilities Plan. The board may now, at the request of a developer, override a local rejection of a proposed hazardous waste facility if the facility has received a permit from the state Department of Environmental Resources.

10. For a discussion of the conceptual work that led to the Massachusetts act, see Michael O'Hare, Lawrence Bacow, and Debra Sanderson, *Facility Siting and Public Opposition* (New York: Van Nostrand Reinhold, 1983); on dispute resolution, see for example, Gail Bingham, *Resolving Environmental Disputes: A Decade of Experience* (Washington, D.C.: Conservation Foundation, 1986).

11. Massachusetts Hazardous Waste Facility Siting Act, Chapter 508, Acts of 1980 (M.G.L. Ch. 21D).

12. Roger Kasperson has dubbed the Massachusetts approach "bartered consent" and has noted that "although the bartered consent model envisions a community social dynamic geared to growing community consensus on the terms necessary for residents to *accept* a facility, the dynamic that actually occurs is one that shows the classic features of social protest: local efforts to mobilize social resources and to identify institutional opportunities by which to *resist* the facility" (emphasis in the original). Roger E. Kasperson, "Hazardous Waste Facility Siting: Community, Firm, and Governmental Perspectives," in *Hazards: Technology and Fairness* (Washington, D.C.: National Academy Press, 1986), p. 137.

13. Massachusetts Senate Bill no. 1073, 1986.

14. New Jersey Major Hazardous Waste Facilities Siting Act, S-1300, P.L. 1981, c. 279 (N.J.S.A. 13:1E–49, *et seq.*).

15. The following discussion of the New Jersey approach is based on the *New Jersey Hazardous Waste Facilities Plan*, March 1985, on other public information material supplied by the staff of the New Jersey Hazardous Waste Facilities Siting Commission, and on extensive conversations with Susan B. Boyle, assistant director of the commission.

16. The act also precludes any major hazardous waste facility from being sited within twenty miles of a nuclear power plant at which spent fuel rods are stored on-site. This stipulation was apparently politically induced, and amendments have been proposed recently to either delete it or narrow its geographic scope.

17. New Jersey is currently a net exporter of hazardous waste; in 1983, for example, New Jersey generators manifested 225,000 tons to other states (primarily to New York, Ohio, and Pennsylvania, for disposal in landfills or for combustion as an auxiliary fuel in cement kilns), whereas New Jersey received approximately 165,000 tons from other states (primarily for aqueous treatment or for incineration).

18. The MSSS grant may not be used for other purposes, such as attorneys' fees.

19. This could be interpreted as a possible justification for the Massachusetts approach. However, it may simply be grounded in a rejection of the approach at hand.

20. The former is envisioned in the siting act, but the latter is questionable legally. Even if it were possible, it is doubtful whether, given the prevailing antagonism toward hazardous waste facilities, a developer would risk investing in a facility that could be closed solely by a local decision.

Annotated Bibliography _____

ASSEMBLED BY **BRUCE PIASECKI**

ABSTRACTS OF RELATED ARTICLES AND RESEARCH EFFORTS

The references listed below are grouped under five subheadings: (1) general information on facility siting; (2) national and state estimates of hazardous waste generation and management capacity; (3) usefulness of compensation, incentives, and negotiation in facility siting decisions; (4) hazardous waste source reduction strategies (annotated); and (5) a general section reviewing some of the most important related books and articles in the field.

Many of the references from subheadings 1 through 3 were selected from a more extensive bibliography prepared for the Council of Planning Librarians by Susan B. Boyle, Associate Director of New Jersey's Hazardous Waste Facilities Siting Commission. That bibliography contains a number of references on the siting of other types of facilities (for example, nuclear power plants and radioactive waste disposal sites). The citation is listed below; copies can be ordered directly from CPL Bibliographies, 1313 East 60th Street, Chicago, Illinois, 60637.

The newest additions to the first three subheadings are the state estimates of generation and management capacity. Many of these plans have been published since 1982. Copies can be ordered directly from each state's hazardous waste siting authority. A directory of these groups is available from the Consortium of State Hazardous Waste Siting Authorities, Inc., 60 West Street, Suite 200, Annapolis, Maryland, 21401. Subheading 4 was also assembled by Susan Boyle and selected and edited by Bruce Piasecki. The final review of related books was assembled and edited by Bruce Piasecki, although written by his colleagues whenever stated after each entry.

1. General Information on Facility Siting

Anderson, Richard F., and Michael R. Greenberg. "Hazardous Waste Facility Siting: A Role for Planners." *Journal of the American Planning Association* 48 (Spring 1982): 204–18.

Booz, Allen & Hamilton, Inc., and Environmental Resources Management. *Guidelines for Siting Hazardous Waste Management Facilities.* Annapolis: Maryland Environmental Service, 1981.

Boyle, Susan B. "Municipal Involvement Is Crucial for Siting New Hazardous Waste Management Facilities in New Jersey." *Public Works* 115 (May 1984): 73–74.

Boyle, Susan B. *Siting New Hazardous Waste Management Facilities through a Compensation and Incentives Approach: A Bibliography.* Chicago: CPL Bibliographies, 1983.

Buckingham, Phillip K., and Richard J. Gimello. "Facility Siting—Getting There in Spite of It All." Paper delivered at the Second Annual Hazardous Materials Management Conference, Philadelphia, Pennsylvania, June 5–7, 1984.

Centaur Associates, Inc. *Siting of Hazardous Waste Management Facilities and Public Opposition.* Washington, D.C.: U.S. Environmental Protection Agency, 1979.

Clark-McGlennon Associates. *Criteria for Evaluating Sites for Waste Management.* Boston: New England Regional Commission, 1980.

Clark-McGlennon Associates. *A Decision Guide for Siting Acceptable Hazardous Waste Facilities in New England.* Boston: New England Regional Commission, 1980.

Council of State Governments. *Waste Management in the States.* Lexington, Mass.: The Council of State Governments, 1982.

Danne, William H., Jr. "Applicability of Zoning Regulations to Waste Disposal Facilities of State or Local Governmental Entities." *American Law Reports* 3d ser., 59 (1983): 1244–75.

Dodd, Frank J. "Siting Hazardous Waste Facilities in New Jersey: Keeping the Debate Open." *Seton Hall Legislative Journal,* 9 (Spring 1986): 423–436.

Dubert, John A., Michael L. Frankel, and Christopher L. Niemczewski. "Siting of Hazardous Waste Management Facilities and Public Opposition." *Environmental Impact Assessment Review* 1 (March 1980): 84–88.

Environmental Resources Management, Inc. *Technical Criteria for Identification and Screening of Sites for Hazardous Waste Facilities.* Trenton: Delaware River Basin Commission and New Jersey Department of Environmental Protection, 1981.

Farkas, Alan L. "Overcoming Public Opposition to the Establishment of New Hazardous Waste Disposal Sites." *Capitol University Law Review* 9 (Spring 1980): 451–65.

Goldshore, Lewis. "Hazardous Waste Facility Siting." *New Jersey Law Journal* 108 (November 19, 1981): 453.

Greenberg, Michael R., and Richard F. Anderson. *Hazardous Waste Sites: The Credibility Gap.* New Brunswick, N.J.: Center for Urban Policy Research, 1984.

"Hazardous Facilities and Land Use Policy." *American Land* 1 (Autumn 1980): 13–22.

Hazardous Waste Dialogue Group (Conservation Foundation). *Siting Hazardous Waste Management Facilities.* Washington, D.C.: Conservation Foundation, 1983.

Hendrickson, Mark L., and Stephen A. Romano. "Citizen Involvement in Waste Facility Siting." *Public Works* 113 (May 1982): 76–79.

Hurley, Mike. *Social and Economic Issues in Siting a Hazardous Waste Facility: Ideas for Communities and Local Assessment Committees.* Fall River, Mass.: Citizens for Citizens, Inc., 1982.

Jaffe, Martin. "Deadly Gardens, Deadly Fruit." *Planning* 47 (April 1981): 16–18.

Kelmas, Deborah, and William Guthe. "Siting Hazardous Waste Facilities in New Jersey Using a Geographic Information System." Paper delivered at the Annual Meeting of the Association of American Geographers, Detroit, Michigan, April 24, 1985.

Kovalick, Walter W., Jr., ed. *State Decision-Makers' Guide for Hazardous Waste Management*. Washington, D.C.: U.S. Environmental Protection Agency, 1977.

Lanard, James S. "The Major Hazardous Waste Facilities Siting Act." *Seton Hall Legislative Journal* 6 (Winter 1983): 367–87.

McAvoy, James F. "Hazardous Waste Management in Ohio: The Problem of Siting." *Capital University Law Review* 9 (Spring 1980): 435–50.

Morell, David, and Christopher Magorian. *Siting Hazardous Waste Facilities: Local Opposition and the Myth of Preemption*. Cambridge, Mass.: Ballinger Publishing Company, 1982.

Mumphrey, Anthony J., Jr., John E. Seley, and Julian Wolpert. "A Decision Model for Locating Controversial Facilities." *Journal of the American Institute of Planners* 37 (November 1971): 397–402.

National Conference of State Legislatures. *Hazardous Waste Management: A Survey of State Legislation, 1982*. Denver: National Conference of State Legislatures, 1982.

National Governors Association. *Review of Sixteen State Siting Laws*. Washington, D.C.: National Governors' Association, 1980.

O'Hare, Michael, Lawrence Bacow, and Debra Sanderson. *Facility Siting and Public Opposition*. New York: Van Nostrand Reinhold, 1983.

Popper, Frank J. "Siting LULUs." *Planning* 47 (April 1981): 12–15.

Rogers, Golden and Halpern. *New Jersey Hazardous Waste Facility Site Search—Interim Report*. Trenton: New Jersey Hazardous Waste Facilities Siting Commission, 1986.

Sandman, Peter M. *Getting to Maybe: Some Communication Aspects of Siting Hazardous Waste Facilities*. Trenton: New Jersey Hazardous Waste Facilities Siting Commission, 1985.

Waldrop, Philip. *State Hazardous Waste Regulations and Legislation: A Synopsis of Information on Seven Selected States*. Washington, D.C.: U.S. Environmental Protection Agency, 1976.

Wetmore, Robert D. "'Massachusetts' Innovative Process for Siting Hazardous Waste Facilities." *Environmental Impact Assessment Review* 1 (June 1980): 182–84.

Wolf, Sidney M. "Public Opposition to Hazardous Waste Sites: The Self-Defeating Approach to National Hazardous Waste Control under Subtitle C of the Resource Conservation and Recovery Act of 1976." *Boston College Environmental Affairs Law Review* 8 (1980): 463–540.

2. National and State Estimates of Hazardous Waste Generation and Management Capacity

Arthur D. Little, Inc. *Hazardous Waste Quantities and Facility Needs in Maryland*. Annapolis: Maryland Hazardous Waste Facilities Siting Board, 1981.

Booz, Allen & Hamilton, Inc. *Executive Summary, Options for Establishing Hazardous Waste Management Facilities: Technology for Managing Hazardous Waste*. Albany, N.Y.: New York State Environmental Facilities Corporation, 1979.

Booz, Allen & Hamilton, Inc. *Options for Establishing Hazardous Waste Management Facilities*. Albany, N.Y.: New York State Environmental Facilities Corporation, 1979.

Busmann, T., et al. *New Jersey Industrial Waste Study: Waste Projection and Treatment.* Washington, D.C.: U.S. Environmental Protection Agency, 1984.

Chemical Manufacturers Association. *The Chemical Manufacturers Association Hazardous Waste Survey for 1981 and 1982.* Washington, D.C.: Chemical Manufacturers Association, 1983.

Chemical Manufacturers Association. *Results of the 1984 CMA Hazardous Waste Survey.* Washington, D.C.: Chemical Manufacturers Association, 1986.

Connecticut Hazardous Waste Management Service. *Connecticut Hazardous Waste Management Plan (Draft)* (and accompanying documents). Hartford: Connecticut Hazardous Waste Management Service, 1985.

Environmental Resources Management, Inc. *Comprehensive Hazardous Waste Management Study.* Jackson: Mississippi Hazardous Waste Council, 1982.

Environmental Resources Management, Inc. *Pennsylvania Hazardous Waste Facilities Plan (Draft).* Harrisburg: Pennsylvania Department of Environmental Resources, 1985.

Hazardous Waste Disposal Advisory Committee. *A Comprehensive Program for Hazardous Waste Disposal in New York State.* 2 vols. Report Number SW-P-16. Albany, N.Y.: Hazardous Waste Disposal Advisory Committee, March 1980.

Illinois Environmental Protection Agency. *Annual Report on Hazardous Waste.* Springfield: Illinois Environmental Protection Agency, 1984.

Kentucky Division of Waste Management. *Generation of Hazardous Waste in Kentucky.* Frankfort: Natural Resources and Environmental Protection Cabinet, 1984.

Maryland Hazardous Waste Facilities Siting Board. *Hazardous and Industrial Waste Management Services Directory.* Annapolis: State of Maryland Hazardous Waste Facilities Siting Board, 1983.

Massachusetts Department of Environmental Management. *Hazardous Waste Management in Massachusetts.* Boston: Massachusetts Department of Environmental Management, 1982.

Minnesota Waste Management Board. *Estimate of Need (Draft).* Crystal: Minnesota Waste Management Board, 1985.

Minnesota Waste Management Board. *Hazardous Waste Management Plan (Revised Draft).* Crystal: Minnesota Waste Management Board, 1984.

New England Congressional Institute. *Hazardous Waste Management in New England.* Washington, D.C.: New England Congressional Institute, 1986.

New Jersey Hazardous Waste Facilities Siting Commission. *New Jersey Hazardous Waste Facilities Plan.* Trenton: New Jersey Hazardous Waste Facilities Siting Commission, 1985.

New York State Department of Environmental Conservation. *Hazardous Waste Disposal Sites in New York State.* 1st Annual Report. Albany, N.Y.: Department of Environmental Conservation, 1980.

New York State Hazardous Waste Treatment Facilities Task Force. *Final Report (Draft).* Albany: New York State Hazardous Waste Treatment Facilities Task Force, 1985.

Radian Corporation. *Hazardous Waste Treatment in North Carolina: A Comprehensive Plan.* Raleigh: North Carolina Hazardous Waste Treatment Commission, 1985.

Southern California Hazardous Waste Management Project. *Hazardous Waste Generation and Facility Development Requirements in Southern California: A Policy Analysis.* Los Angeles: Southern California Hazardous Waste Management Project, 1983.

Westat, Inc. *National Survey of Hazardous Waste Generators and Treatment, Storage, and Disposal Facilities Regulated under RCRA in 1981.* Washington, D.C.: U.S. Environmental Protection Agency, 1984.

3. Usefulness of Compensation, Incentives, and Negotiation In Facility Siting Decisions

Bacow, Lawrence R., and Debra R. Sanderson. *Practical Guidelines for Negotiating Compensation Agreements.* Cambridge, Mass.: Massachusetts Institute of Technology Energy Impacts Project, 1980.

Clark-McGlennon Associates. *Negotiating to Protect Your Interests.* Boston: New England Regional Commission, 1980.

Kretzmer, David. *Binding Communities to Compensation Agreements for Facilities.* Cambridge, Mass.: Massachusetts Institute of Technology Energy Impacts Project, 1979.

Kunreuther, Howard. "Balancing Risks and Benefits in Siting Hazardous Facilities." *Bell Atlantic Quarterly* 2 (Winter 1985); pp. 37–48.

Mumphrey, Anthony J., Jr., and Julian Wolpert. "Equity Considerations and Concessions in the Siting of Public Facilities." *Economic Geography* 49 (April 1973): 109–21.

O'Hare, Michael. " 'Not on *My* Block You Don't': Facility Siting and the Strategic Importance of Compensation." *Public Policy* 25 (Fall 1977): 407–58.

Rivkin, Malcolm D. *An Issue Report: Negotiated Development: A Breakthrough in Environmental Controversies.* Washington, D.C.: Conservation Foundation, 1979.

Sanderson, Debra R. *Compensation in Facility Siting Conflicts.* Cambridge, Mass.: Massachusetts Institute of Technology Energy Impacts Project, 1979.

Sanderson, Debra R. *Facility Siting, Social Costs, and Public Conflict.* Cambridge, Mass.: Massachusetts Institute of Technology Energy Impacts Project, 1979.

Smith, Martin, et al. *Costs and Benefits to Local Government Due to Presence of a Hazardous Waste Management Facility and Related Compensation Issues.* Chapel Hill: Institute for Environmental Studies, 1985.

U.S. Environmental Protection Agency. *Using Compensation and Incentives When Siting Hazardous Waste Management Facilities.* SW–942. Washington, D.C.: U.S. Environmental Protection Agency, 1982.

U.S. Environmental Protection Agency. *Using Mediation When Siting Hazardous Waste Management Facilities.* SW–944. Washington, D.C.: U.S. Environmental Protection Agency, 1982.

4. Hazardous Waste Source Reduction Strategies

Adamson, Virginia F., *Breaking the Barriers: A Study of Legislative and Economic Barriers to Industrial Waste Reduction and Recycling.* Toronto, Canada: Joint Project between the Canadian Environmental Law Research Foundation and the Pollution Probe Foundation, 1984.

This source examines the trend toward low-waste technologies and management methods despite often conflicting regulations and financial disincentives. Adamson presents solid information on the costs of capital investments, disposal, and transportation. The study discusses Canada's various loan programs and indirect subsidies while also reviewing the problem created by insufficient information. The recommendations are detailed, thoughtful, and fully developed for both industry and government.

Banning, W., and S.H. Hoefer, for Manufacturers Association of Central New York, Central New York Regional Planning and Development Board, and New York State Environmental Facilities Corporation. *An Assessment of the Effectiveness of the Northeast Industrial Waste Exchange in 1982.* February 1983.

Two primary functions are evaluated to measure the success of the Northeast Industrial Waste Exchange during 1982: (1) its effectiveness in creating a market for waste materials, and (2) its effectiveness in achieving successful exchanges between companies.

Battelle (Columbus Division) for New York State Environmental Facilities Corporation. *A Preliminary Handbook on the Potential of Recycling or Recovery of Industrial Hazardous Wastes in New York State.* July 1982.

The potential of recycling or recovering 47 types of hazardous wastes generated by industries under 17 different SIC categories is evaluated. Identification of waste types is based on the New York State Department of Environmental Conservation (NYSDEC) inventory of industrial hazardous wastes. For each waste type, information is given on probable characteristics of waste, potential recycling/recovery techniques, and potential value of recycled/recovered waste materials or by-products. Industrial groups included are industrial inorganic chemicals (SIC 281), plastic materials, synthetic resins and rubber, and man-made fiber (SIC 282), and coating, engraving, and allied services (including job-shop electroplating) (SIC 347).

Cole, C. A. *State of Technology in Recycle, Reuse, and Recovery of Hazardous Waste in the Plastics and Synthetics Industries in Pennsylvania.* University Park, Pa.: Pennsylvania State University, November 1984.

The report describes state-of-the-art technologies used by the plastics and synthetics industries in Pennsylvania to treat hazardous and residual wastes. It includes a survey of 29 Pennsylvania industries to determine their problems and research needs. Survey results indicated that industry perceives its problems as minor at this time; the worst problems are related to permitting and public opinion. Product lines, quantities of generated waste, and existing recycle reuse, recovery, treatment, and disposal options are listed. Various waste control options are examined with respect to their advantages, disadvantages, and current usage. An index of documents and organizations is included that supplies additional detailed information on the report's topic.

Environmental Resources Management, Inc., for New York State Department of Environmental Conservation. *Hazardous Waste Technology Study.* 1984.

This report reviews hazardous waste management technologies for waste reduction, recovery, and treatment. Waste reduction includes process modification, chemical substitution, and on-site recycling and treatment. Recovery technologies include waste exchange, recovery of organic constituents, acid regeneration, and energy recovery. Treatment technologies include thermal destruction and chemical/physical/biological treatment. An overview of technologies shows their applicability to the prevalent types of waste generated in New York State.

Herndon, R. C., and E. D. Purdum, eds., for the U.S. Environmental Protection Agency and Florida State University. *The Second National Conference of Waste Exchange.*

March 1985.

Proceedings of this conference include separate sessions on waste exchanges in North America, waste exchange for high-tech industries, and waste exchange for small-quantity generators. Relevant legislation and regulations were presented in other sessions of the conference and were supplemented with perspectives on waste exchanges.

Huisingh, D., H. Hilger, S. Thesen, and L. Martin, for the North Carolina Board of Science and Technology. *Profits of Pollution Prevention—A Compendium of North Carolina Case Studies in Resource Conservation and Waste Reduction*. May 1985.

This compendium presents 25 case studies organized by industry groups, including four case studies in the chemicals industry, two case studies in "high technologies" industries, and three case studies in metal plating/working industries. Each case study describes a project for resource conservation and/or waste reduction, the environmental and health benefits of carrying out the project, and resultant cost savings, if any. All projects are evaluated by standardized parameters.

ICF Consulting Associates, Inc. *Overview of Incentives for and Barriers to Hazardous Waste Reduction in the State of California*. Los Angeles: California Department of Health Services, June 1985.

ICF reviews the technical and economic barriers to hazardous waste reduction. SICs 2816, 2821, 3471, and 3679 are four of nine industries targeted. ICF outlines each industry's characteristics (number of plants, number of employees, sales, expected growth, and so on) and assesses the ability of each to incorporate new technologies. Technological case studies are not presented, but available and proven hazardous waste reduction technologies in each industry are identified. For example, in SIC 2821, operations listed for solvent recovery include filtration, air stripping, carbon adsorption, and ion exchange. The financial analysis given provides good information on the critical elements involved in securing financial assistance.

ICF Consulting Associates, Inc. *Review of Incentive Programs in Selected States and Canada*. Los Angeles: California Department of Health Services, 1985.

This review identifies the different types of incentive programs being implemented by government to encourage source reduction among industries. Different programs generally reflect different goals of the agency, needs of the industries, resources, and geographic layout. Careful consideration is given to four main topic areas: technical assistance, grants, loans, and tax incentives. An analysis is presented on the effectiveness of centralized versus decentralized government involvement.

Jacobs Engineering Group, Inc., for the U.S. Environmental Protection Agency. *Waste Minimization*, Vol. 2, *Source Reduction*. October 1985.

This report presents a national profile of source reduction activities based on studies by industrial processes. Process and practice studies include SIC 2816 (inorganic pigments), SIC 2822 (synthetic rubber manufacture), SIC 3471 (electroplating/metal surface treatment), and SIC 3679 (printed circuit boards). Incentives/disincentives relating to source reduction have also been evaluated.

JRB Associates, for the Maryland Hazardous Waste Facilities Siting Board. *Three Case Studies to Improve Industrial Waste Management in Baltimore-Area Metal-Finishing Plants*. February 1984.

The metal-finishing industry (SIC 34, 35, and 36) was evaluated in this report. Case studies include a job shop, a captive shop, and a circuit-board manufacturer. Recommendations were developed to reduce the quantity of hazardous waste generation at metal-finishing plants and disposal centers. The results of the implemented recommendations were also evaluated.

Kohl, J., and B. Triplett. *Managing and Minimizing Hazardous Waste Metal Sludges: North Carolina Case Studies, Services, and Regulations.* Manual prepared with support from the Governor's Waste Management Board and the North Carolina Hazardous Waste Management Branch at Raleigh, North Carolina. December 1984.

This manual for managing electroplating sludge presents case studies of companies that have installed metal recovery equipment, reduced their metals consumption, and taken other steps to reduce their hazardous waste sludge problems. Six case studies of source reduction techniques are described.

Kohl, J., P. Moses, and B. Triplett. *Managing and Recycling Solvents—North Carolina Practices, Facilities, and Regulations.* Manual prepared with support from the North Carolina Board of Science and Technology, Raleigh, North Carolina. December 1984.

This manual examines current practices and alternatives for managing waste solvents in North Carolina industry. Procedures for solvent management are discussed. Both vapor (or hot) degreasing and liquid (or cold) degreasing are studied. Studies on product-finishing, coating, printing, and wood furniture industries identify alternatives for current solvent management techniques. Twenty-six case studies of source reduction techniques are described.

Lancy, L. E. *Waste Treatment and Hazardous Waste Management Problems Facing the Metal Finishing Industry in Pennsylvania.* University Park, Pa.: Pennsylvania State University, October, 1984.

This report examines state-of-the-art technologies available for treating solid and liquid wastes from the metal-finishing industry. It also describes the waste control options available primarily to job shops, which are threatened by economic constraints. The industry, the nature of the wastes requiring treatment, and the applicable available treatment technologies are briefly discussed. Discharge limitations for the metal-finishing industry, justification for technical assistance to the electroplating job shop, and treatment opportunities currently available for job shops are presented. Recommendations outlining how cooperative efforts between the state and educational institutions could assist electroplaters is also provided. To complement the broad outlines of the report, references for detailed data are supplied.

League of Women Voters of Massachusetts. *Waste Reduction: The Untold Story.* Conference held at the National Academy of Sciences Conference Center at Woods Hole, Massachusetts, June 19–21, 1985.

This collection of questionnaires describes the source reduction accomplishments of 23 major U.S. corporations. Questionnaires requested information on net worth, number of employees, sales, major products, number of plants, major hazardous wastes, corporate policies, and hazardous waste reduction accomplishments over the past five years. Companies in SICs 2819 and 2821 are well represented. SICs 3471 and 3679 are represented to a lesser degree.

Marlow, H. LeRoy. *Another Perspective on Technology Transfer: The PennTAP Expe-*

rience. University Park, Pa.: Pennsylvania State University, 21 May 1984.

In this report, Marlow discusses Pennsylvania's philosophy of technology transfer. Four specific transfer philosophies have evolved in this program, which the author feels should be considered for any similar program. Marlow concludes that information dissemination is a human function and that all research data must be translated into a language all can understand. He also points out that the availability of information does not ensure its use. But of equal importance, Marlow advocates avoiding transfer technologies that merely reinvent the wheel. Moreover, PennTAP's specific objectives are reviewed, and the characteristics of its program are presented.

Marlow, H. LeRoy. "Technology Transfer—The State's Point of View." Paper presented at the U.S. Atomic Energy Commission Technology Transfer Meeting, Washington, D.C., 8 March 1973.

This paper discusses the state's view of technology transfer as it pertains to cooperation with federal agencies. Technology transfer is defined as the entire process whereby existing technical data and expertise is applied to meet the scientific and technical needs of the public or private sector. Several state technology transfer systems are briefly reviewed. The state methods use information depositories, educational activities, and field agents to utilize technology transfer. These are discussed as well. The Pennsylvania Technical Assistance Program is also used to illustrate successful technology transfer by a state. Possible state/federal areas of cooperation are identified.

Massachusetts Department of Environmental Management. *Hazardous Waste Source Reduction Sessions*. Boston, Massachusetts, 1984/85.

The Massachusetts DEM held several source reduction sessions for specific waste streams such as metallic wastes, solvent wastes, and electroplating wastes. Sessions were conducted by experienced consultants who reviewed applicable regulations and assessed major generating industries. The consultant also outlined those wastes most frequently generated and major factors affecting reduction. The proceedings of each session provide a list of recovery techniques and equipment use by industry. Industry's success with these techniques and equipment is also documented.

Minnesota Mining and Manufacturing Company (3M Company) for the United Nations Environment Programme, Office of Industry and the Environment. *Low or Non-Pollution Technology through Pollution Prevention,* 1984.

This booklet presents a perspective on industrial pollution problems, concepts and principles, cost/benefit evaluation, and examples of low and non-pollution technology.

Minnesota Technical Assistance Program. *Fact Sheets/MnTAP Services Packet*. Minnesota Waste Management Board.

MnTAP determined that generators needed a concise, easy-to-read sheet of technical information on waste examples and sources, waste management, hazardous characteristics, and waste reduction options for various waste streams. Technical information on acids, bases, solvents, paints, PCBs, waste oil, printing inks, lab wastes, metal sludges, and pesticides will be included on each sheet.

National Research Council. *Reducing Hazardous Waste Generation*. Washington, D.C.: National Academy Press, 1985.

This general information source on waste reduction examines the varied and

complex factors affecting industry's decisions on waste reduction, including economics, technology, regulatory climate, and attitudes. It concludes that decisions
to reduce waste tend to be specific to the industry and often to the plants, and
may be largely influenced by non-technological factors. This source presents a
framework for considering the myriad possible non-technological factors and approaches for encouraging hazardous waste reduction.

New Jersey Department of Environmental Protection, Division of Waste Management.
   Source Reduction of Hazardous Waste. Seminar presented by the Division of
   Waste Management's Program at Douglas College, Rutgers University, 22 August
   1985.

   Proceedings of this seminar provide insight to source reduction programs being
   developed in the industry, including case studies from 3M Company, Allied
   Corporation, Pioneer Metal Finishing, Inc., and organic chemical manufacturing
   plants in New Jersey. Emerging technologies of waste management have been
   reviewed in one of the papers. Another paper examines key institutional, non-
   technical factors that affect the generation of hazardous waste in industry.

New Jersey Hazardous Waste Facilities Siting Commission et al. *Proceedings of the New
   Jersey Hazardous Waste Sources Reduction and Recycling Roundtable.* Conference held in Princeton, New Jersey, 25 July 1984.

   These proceedings contain useful case studies from 3M Company, Allied Corporation, Shell Oil Company, and Union Carbide Corporation. 3M's report discusses how all 3M employees are involved in identifying areas of potential
   pollution or waste prevention. The program provides workers with knowledge of
   pollution prevention and from that knowledge, coordinate employee plant inspections that could serve as a foundation for developing innovative source reduction programs. Allied lists several examples of waste reuse. With liabilities
   in mind, Shell Oil describes the benefits of waste reduction and recycling. However, Shell indicates that waste reduction will be limited until costs are comparable
   with less expensive treatment and disposal options. Union Carbide presents its
   approach to waste reuse and reduction. The strategy is motivated by cost avoidance.

Norgaard, M.A., for the Minnesota Waste Management Board's Waste Reduction Conference. *Waste Reduction—A Catalog of Opportunities.* September 1983.

   Some of industry's waste reduction opportunities are described. This catalog
   outlines and serves as a guide for specific waste reduction practices. Waste reduction opportunities for metal-finishing processes are included.

North Carolina Department of Natural Resources and Community Development. *Accomplishments of North Carolina Industries.* Raleigh, North Carolina: Pollution Prevention Pays Program, 1984.

   This compendium depicts North Carolina industries' source reduction programs
   and includes a listing of individual companies, a description of the source reduction
   efforts, and the savings involved. The compendium is comprised of recipients of
   the North Carolina Governor's Award for Excellence in Waste Management.

North Carolina Department of Natural Resources and Community Development. *Pollution
   Prevention Bibliography.* Compiled under the Pollution Prevention Pays Program
   at Raleigh, North Carolina. January 1985.

   This bibliography provides specific information about chemical and allied products, fabricated metal products (except machinery and transportation equipment),
   and electronic equipment and supplies.

North Carolina Department of Natural Resources and Community Development. *Pollution Prevention Challenge Grants,* 1985.

This pamphlet describes financial assistance provided in the form of matching grants to businesses and communities throughout North Carolina under the Pollution Prevention Pays Program. The following are identified: purpose of the challenge grants, eligibility for grants, project focus, funding and timing, evaluation criteria, grant proposal forms and content, nature of grant agreement, and the agency contacts.

Ontario Research Foundation, Proctor & Redfern Ltd., and Weston Designers, for Ontario Waste Management Corporation. *Waste Reduction Opportunities Study.* January 1983.

This study involved a survey of three major industry sectors: primary metals, fabricated metal parts, and chemical manufacturing. For each of the industry sectors, approximately 100 companies were selected. They were contacted first by mail, followed by telephone interviews.

Regan, R. W., and P. E. Craffey, for the Pennsylvania State University. *Assessment of the Hazardous Waste Practices in the Paint and Allied Products Industries.* University Park, Pa.: Pennsylvania State University, December 1984.

This report uses information from technical literature and incorporates data from representatives of the Pennsylvania paint industry. The report addresses the technical alternatives available to the paint industry for the management of small quantities of hazardous and toxic wastes and the costs and legal ramifications of the alternatives. In addition, the report includes recommendations for future actions. Suggestions are provided and could be used by firms with limited technical assistance. A list of references is included.

Smith, Martin A., Research Associate, Institute for Environmental Studies, University of North Carolina at Chapel Hill. *A Handbook of Environmental Auditing Practices and Prospectives in North Carolina.* 1985.

This handbook includes examples of environmental auditing and related programs in North Carolina. Case studies on Duke Power Company, ITT Corporation, the city of Raleigh, and Texas Gulf Chemicals Company are described. Guidelines for hazardous waste management and pollution prevention are developed. The author provides helpful reference materials and sample audit checklists.

Snow, H. "New York State Environmental Facilities Corporation's Program on Assistance to Industry for Waste Reduction." Paper presented at NCASI's Northeast Regional Meeting at Boston, Massachusetts, November 1, 1984.

This report describes the New York State EFC program, where waste generators are encouraged to reduce, recover, and recycle material. The program's industrial clients include AIRCO, General Electric, Lehigh Portland Cement, Brown Company, IBM, Eastern Milk Producers Cooperative, Jamaica Water District, and Crown Zellerbach. Projects focus on pollution control and water and solid waste financing.

University of Maryland. "Opportunities for Industrial Resource Recovery" and "Metals Recovery from Electroplating Wastewaters." *In Industrial Waste: Proceedings of the Fourteenth Mid-Atlantic Conference.* Ann Arbor, Michigan: Ann Arbor Science Publishers, 1982, pp. 117–124 and 125–131.

Both of these articles discuss methods for removing and recovering metals from industrial wastes. Techniques discussed include ion exchange, reverse osmosis, evaporation, electrolytic recovery, electrodialysis, Donnan dialysis, and insoluble starch xanthate. Each technology is compared by chemical costs, energy consumption, and treatable concentrations of wastes.

U.S. Congress. Office of Technology Assessment. *Technologies and Management Strategies for Hazardous Waste Control*. March 1983.

OTA identifies four policy options that lay the foundation for an immediate and comprehensive approach to protecting human health and the environment from the dangers posed by mismanagement of hazardous waste. The report also identifies key issues and summarizes OTA's evaluation of the current Federal program under RCRA and CERCLA.

U.S. Environmental Protection Agency. *Control and Treatment Technology for the Metal Finishing Industry—Ion Exchange*. Summary report developed by the Industrial Environmental Research Laboratory, Cincinnati, Ohio, June 1981.

This summary promotes the use of ion exchange in the metal-finishing industry. Basic ion-exchange concepts are discussed. Three major areas of application are considered: wastewater purification and recycle, end-of-the-pipe pollution control, and chemical recovery. These applications are evaluated in terms of performance, state of development, cost, and operating reliability.

U.S. Environmental Protection Agency. *EPA Alternative Treatment Studies*. Office of Solid Waste, Waste Management and Economics Division, Washington, D.C., April 1985.

This catalog reviews EPA's alternative treatment studies started by the Office of Solid Waste (OSW) during 1985. This project includes studies on waste minimization, macroscale assessment of technology capacity, and waste stream characterization. The costs and time to develop technologies are also estimated. Lists of contacts at OSW and lists of contractors are included.

U.S. General Accounting Office. *State Experiences with Taxes on Generators or Disposers of Hazardous Waste*. Report to the Chairman, Subcommittee on Commerce, Transportation and Tourism, Committee on Energy and Commerce, House of Representatives. May 1984.

This report discusses the experiences New York, New Hampshire, and California have had with taxes on hazardous waste generation, transportation, treatment, storage and disposal. GAO found that the three states have not collected the revenues they anticipated; have not determined if the tax achieved its objective of discouraging non-desirable waste management practices; and were concerned that a similar federal tax may reduce state tax revenue or increase the incentive to illegally dispose of hazardous waste.

Versar, Inc., for U.S. Environmental Protection Agency, Office of Solid Waste, Waste Treatment Branch. *Waste Minimization*. Vol. 1, *Issues and Options*. October 1985.

The report presents an assessment of government, industry, non-governmental, or non-industrial efforts to promote waste minimization. Regulatory options with or without additional legislation, as well as non-regulatory options, have been evaluated.

Versar, Inc., for U.S. Environmental Protection Agency, Office of Solid Waste, Waste Treatment Branch. *Waste Minimization*. Vol. 3, *Recycling Practices, Incentives, and Constraints*. October 1985.

This report presents a profile of specific trends and incentives/disincentives for recycling five generic hazardous waste stream categories: solvents, halogenated organics other than solvents, metals, corrosives, and cyanides/other reactives.

Walters, R.W., for the Hazardous Waste Facilities Siting Board. *Opportunities for Technical Assistance with Industrial and Hazardous Waste Management for Maryland Industries*. January 1984.

This study identifies available academic expertise pertinent to waste management. In addition, an industrial survey to determine specific waste management problems is described. Three case studies are used to demonstrate that the industry could benefit from a technical assistance program. Case studies included Solvent Recovery in a Paint-Coating Facility; Electrolytic Recovery of Zinc from Metal Finishing Rinse Water; and Metal Precipitation from Pickle Liquor.

Abstracts of Related Books and Articles

The following five papers were presented at the First International Symposium on Operating European Centralized Hazardous (Chemical) Waste Management Facilities. This important event was held in Odense, Denmark, on September 20–23, 1982, and signified Europe's early recognition of the need for strong government leadership in planning for safe waste treatment. The five abstracts were written by Bill Armbruster, a student of Bruce Piasecki's at Clarkson University.

a. The Danish System

Speaker: Peter Løvgren, Managing Director, ChemControl A/S

As noted throughout *America's Future in Toxic Waste Management,* Denmark is a pioneering country in hazardous waste management. The types of waste generated are similar to those of other industrialized countries. The Danish system manages 51 different categories of hazardous waste. The paper discusses the options available to waste generators in Denmark.

If a permit is granted, waste generators can treat, deposit, or recycle the waste on-site. Wastes can be sent to an approved centralized or regional waste recycling or treatment facility. If permitted, waste generators can sell the waste to another industry or through the waste exchange. Examples of transporting wastes to firms treating waste on-site and transporting wastes to regional recycling facilities are provided. For example, the centralized Danish system is introduced and the unique network of transfer stations is discussed.

A communication network has proven instrumental for controlling Denmark's integrated system. Moreover, communication with the public, especially with the neighbors of the treatment plant, has been regarded as an important means of getting the system accepted.

Since operations began in 1975, five accidents have occurred. While these mishaps are briefly described, the safe and successful operation of the Danish system is emphasized.

b. The Legislation Regarding Oil and Chemical Waste in Denmark

Speaker: Jorgen Lauridsen, M.Sc. (Chem Eng), Chief Engineer, Danish Environmental Protection Agency

This paper describes the administration and supervision of the Danish legislation. The Central Administration of Environmental Affairs consists of one department with five related agencies. The National Agency of Environmental Protection (NAEP) is the central body in the field of pollution control and a part of the Ministry of the Environment since April 1972. The other agencies are the National Agency of Physical Planning; the National Agency of the Protection of Nature, Monuments, and Sites; the National Forest Service, which manages forest and dune areas; and the National Food Institute. The legal authority for promulgating rules is Act no. 178 of May 24, 1972 on "Disposal of Oil Waste and Chemical Waste." A review of the act and its legislation most relevant to toxics is made in this useful paper.

c. Receipt at Kommunekemi; Inorganic Physical/Chemical Treatment Plant

Speaker: Peter Henriksen, M.Sc. (Chem Eng), Kommunekemi A/S

This paper is divided into four parts. The first focuses on practical questions regarding the receipt of wastes. To ensure that wastes are treated safely, different waste streams are delivered separately and their compositions are listed. As a result, every waste delivery to Kommunekemi is accompanied by a manifest that includes the following information: conditions for receiving the waste, type of waste, waste-producing process, quantity and way of delivery, waste producer, and a contact person.

The second section describes the storage facilities. The Kommunekemi plant has six essential storage parts, each of which is described in detail: (1) a reception platform; (2) a plant for emptying railway tank cars and sludge trucks; (3) a plant for emptying drums with fluid waste; (4) a tank farm for storing fluid waste; (5) warehouses for storing packaged solid waste; and (6) in connection with the waste oil plant, a tank farm for storing oil waste and refined oil. The description of these facilities should prove most useful to U.S. firms contemplating similar units.

The third section discusses the processes of the inorganic chemical–processing plant. The main part of the inorganic chemical wastes originate from acid pickling baths, electroplating baths, and sludge from the wastewater plants of electroplating industries. Most of these wastes are either aqueous solutions or products that can be dissolved in water. The processing method includes oxidation of cyanide, the reduction of chromate, and the precipitation of heavy metals as hydroxides.

The last section is on the treatment of wastewater. Kommunekemi does not have a wastewater treatment plant; instead, the wastewater from the inorganic plant and the surface water passes through an observation basin. The basin, however, is equipped with a skimmer to retain any oil. Wastewater is then passed through the basins, where water quantity, temperature, and pH are registered automatically and samples are taken for further management decisions.

d. Rotary Kiln Incinerators for Organic Chemical Wastes

Speaker: Arne Kristensen, Plant Manager, Kommunekemi A/S

Kommunekemi has three incinerators for destroying hazardous waste. The capacities and performance of the incinerators include one small plant for the destruction of liquid sulfur and halogenated waste; one large plant for the destruction of solid and liquid waste with a low content of sulfur and halogens; and one large plant for the destruction of solid and liquid waste containing sulfur and halogens.

The incinerators are Kommunekemi's largest treatment units, with an annual treatment capacity of approximately 80,000 tons. Based on their seven years of operating experience, this report summarizes why the Danish operators feel their approach is benign.

e. Clean-Up of Abandoned Sites

Speaker: Jens Kampmann, Director, National Agency of Environmental Protection

This paper describes the cleanup of Cheminova's 3,000–cubic-meter dump near break-water 42 on Harboore Tange and outlines how mapping toxic chemical dumpsites was put into effect nationwide. The mapwork helps obtain information on the number of dumps where chemical wastes have been deposited or buried. To date, 600 dump sites have been mapped, each of which has been categorized in this report.

f. Economical Evaluation of Different Waste Treatment Processes

This paper was presented by the Association of Finnish Waste Contractors during Helsinki Waste Conference. The Danes feel that the policy for handling and treating waste in Denmark has proved to be effective. They attribute the success to the fact that only well-tested, well-proven, and reliable treatment methods have been applied.

A representative from the manufacturers of each successful technology was asked to outline the total cost of investment and operation for disposing of the waste from three different communities. After a comparison of the processes, mass combustion proved the most advantageous. Although capital intensive, this incorporates possibilities for energy recovery.

g. Public–Private Partnership: New Opportunities for Meeting Social Needs

Harvey Brooks, Lance Liebman, and Corinne S. Shelling, eds. Cambridge, Mass.: Ballinger Publishing Company, 1984. 374 pp. Reviewed by Michael Mueller, Economics Department, Clarkson University.

The book consists of sixteen chapters by different authors. The chapters are grouped into five different parts, with each developing a particular aspect of the major partnership theme.

The first part is a chapter by Harvey Brooks that outlines the book's theme. This is followed by a chapter by Thomas K. McCraw that relates the history of public-private interactions in the United States. The next part gives five views of the most critical components of partnerships. Charles Haar stresses the importance of a coordinating role for the federal government; Ted Kolderie says that solutions need to be tailored correctly for each setting; Jordan Baruch notes that significant private and public entrepreneurship is necessary; Orlando Patterson describes the failures and successes of partnership experiences in Jamaica; and Marc Berdick, Jr., stresses that there are no quick and easy efficiencies.

Part three is a theoretical contribution concerning the new, not-solely-for-profit role required of corporations in the partnership theme. William Baumol explains that government action should help weaken the grip of the invisible hand to free firms to pursue broader goals; Robert Clark analyzes the legal responsibilities of corporate officers in light of non-profit goals; and James Worthy expounds on a management theory for non-profit tasks.

The final part describes several areas, possibilities, and future directions for public-private partnerships. William C. Norris explains the experiences of Control Data Corporation with partnerships (CDC has historically pursued many non-profit goals); Robert Reich describes the necessary interplay of corporate interests and government leadership;

Peter Drucker explains that initiative must come from businesses themselves; James L. Sundquist restates and defends the basic theme by analyzing why privatization alone is not a workable solution, because society has interests that are not possible to express in "the market place"; Franklin Long shows the possibility of public-private partnership for multinational corporations and developing countries. In the concluding part Lance Liebman discusses the tensions and emotional reactions that advocates of public-private partnerships might face in this country based on its intellectual and political traditions.

This book provides a worthwhile exposition of public-private interaction as a solution to social problems. It clearly renounces the simplistic solution of reducing the public sector and increasing privatization. The partnership solution is not proposed as a panacea, because many of the chapters stress its problems and difficulties. While hazardous waste problems are not specifically addressed in the book, the relevance is so clear to the purposes of *America's Future in Toxic Waste Management* that readers familiar with hazardous waste problems will find many useful ideas in this broad and provocative book.

Caldwell, Lynton Keith. *International Environmental Policy.* Dunham, N.C.: Duke University Press, 1986, 367 pp. Reviewed by Elizabeth A. Richert.

In his latest book, *International Environmental Policy,* Lynton Caldwell, a renowned scholar of environmental studies and prime craftsman behind the 1969 National Environmental Policy Act, provides a historical account of global environmentalism and outlines the "emergence of a new configuration of international policy." While Caldwell's book focuses on the accomplishments rather than the shortcomings of international environmental protection, the work is not entirely optimistic. It shows, for instance, the necessity for increasing international controls for whaling and endangered species and helps demonstrate the need for more coordinated programs in the management, transport, and treatment of toxic wastes.

Starting with the 1972 United Nations Conference on the Human Environment in Stockholm, Caldwell traces the development of several institutional structures under the guidance of the U.N. He also describes the active development and influential role of non-governmental organizations' actions on international issues. By depicting these organizations' interactions, Caldwell discloses the structures now in place that contribute to the development and recognition of multinational environmental policy.

Of significance is Caldwell's discussion of "Regional Arrangements" as well as local government and citizen approaches to environmental protection. Here the European Community (EC) serves "as an instructive example of the emergence of environment as a focus for international policy." As in the German states described in Piasecki's *America's Future in Toxic Waste Management,* Caldwell notes that the EC acts as a collective in which community interests, such as marine protection strategies, take precedent over "individual national policies."

As pointed out in this book's last chapter, regional arrangements may thereby have a profound impact in promoting the global goals of environmental protection. "Governments must simultaneously be influenced from within, and induced from without, to work with other governments and international organizations," notes Caldwell. Regional arrangements as studied by Piasecki and Davis may ignite the development of an international policy, particularly in the waste management and disposal area, for the rest of the world to follow.

While the obvious strength of Caldwell's book lies in the chapters on the U.N., regional arrangements, and non-governmental organizations, an equally important project would include a disclosure of the weaknesses of the structures Caldwell clearly describes. Only then will we be able to assess the strength of the progress made toward institutions of global environmentalism.

Goldman, Ben, James A. Hulme, Cameron Johnson, and the Council on Economic Priorities. *Hazardous Waste Management: Reducing the Risk*. Washington, D.C.: Island Press, 1986, 316 pp. Reviewed by Elizabeth A. Richert.

Not many corporations want to be responsible for the threats posed by improper waste disposal. Now there is a way to decrease the liabilities associated with mishandling hazardous waste. The investigation conducted by the Council on Economic Priorities (CEP) provides industry with a "consumer guide" to U.S. waste management facilities. This detailed report helps industries choose a waste disposal company that reduces their risk of creating another Love Canal.

The CEP's vigorous 21–parameter evaluation of the "Big Eight" commercial hazardous waste management companies provides hazardous waste generators, state and local governments, and citizen organizations with a "profile of the strengths and weaknesses of specific companies." The CEP developed report cards for U.S. Pollution Control, Inc. (USPCI), U.S. Ecology, Inc. (USEC), CECOS International, Inc. (CECOS), International Technology Corp. (IT), Environmental Systems Company, Inc. (ENSCO), Rollins Environmental Services, Inc. (RES), IU International Corp. (IU), and Waste Management, Inc. (WMI). The council based its grading on corporate financial liabilities, corporate and public relations, compliance with government regulations, groundwater monitoring, and the status and efficacy of technologies employed. Decisions were also supplemented by using the U.S. Environmental Protection Agency's Hazard Ranking System (HRS), which "compared facilities on the bases of site, management and technology." USPCI, Inc., was awarded the highest grade, "B," while IU International Corp. received a "D."

In tandem with *America's Future in Toxic Waste Management*, the CEP report also reviews state-of-the-art disposal technologies. While the council points out that there are discrepancies in the data concerning hazardous waste transportation accidents and that weaknesses exist in national strategies for monitoring groundwater contamination, it provides a description of some of the "emerging technologies that will shape the future of waste management." The CEP's depiction of mobile incineration units for destroying hazardous waste forcefully echoes Piasecki and Davis's claims in their "Concluding Remarks Section." By focusing its effort on securing alternatives to dumping, this novel consumers' guide should be read by all professionals in the field of high-risk management.

Index _____

322   INDEX

waste oil in, 35, 98, 99, 102, 135,
136, 137–43, 146–47, 149–51; waste
statistics, 90–91, 100, 101–2, 103,
104, 105, 118, 139; ZVSMM facilities, 2, 58, 59–60, 100, 101–2, 121,
171, 172, 173, 174
Williams, Marcia, 216
Wisconsin, collection program in,
213

Yakowitz, Harvey, 161

Zinc, recycling of, 29
Zubiller, Carl Otto, 185
ZVSMM facilities, 2, 58, 121; description of, 59–60, 101–2; siting of,
171, 172, 173, 174; statistics, 100,
101–2

About the Authors _____

BRUCE PIASECKI is Associate Director of the Hazardous Waste and Toxic Substance Research and Management Center at Clarkson University, where he is also a professor in the Center for Liberal Studies. He is the author of *Beyond Dumping* (Quorum, 1984) and numerous articles on environmental management, appearing in such journals as *Science 83*, *Technology Review*, *Business and Society Review*, and the *Washington Monthly*. He is a business development consultant and founder of the American Hazard Control Group, which provides legal, planning, and policy advice on hazardous waste management.

GARY DAVIS is a Research Associate at the University of Tennessee's Center for Energy, Resources, and the Environment where he is affiliated with the Waste Management Research and Education Institute. He is also a practicing environmental attorney based in Knoxville. As a hazardous waste specialist for the state of California, Davis helped establish the list of toxic wastes no longer allowed to be land disposed under federal law.

LEE BRECKENRIDGE is Chief of the Environmental Protection Division and Assistant Attorney General in the Massachusetts Department of the Attorney General. She studied waste oil and waste solvent management in the European Communities on an environmental fellowship from the German Marshall Fund of the United States in 1985.

JANET BROOKS is a Staff Attorney for the Connecticut Fund for the Environment. As a recipient of a German Marshall Fund grant, Ms. Brooks interviewed many of Europe's leading citizen and environmental group leaders. Specializing in the role of citizen rights in hazardous waste reform, Ms. Brooks remains quite interested in worker right-to-know issues.

LYNTON K. CALDWELL is a Distinguished Professor of Political Science at the University of Indiana and Director of an advanced institute in science policy studies. As a prime force behind the establishment of the Environmental Impact Statement (EIS) process, Dr. Caldwell is considered one of America's chief architects of the National Environmental Policy Act. He further secured his renown, throughout the 1970s, with his efforts to refine the EIS process into a quality guidance tool for the revitalization of American industry. The holder of over two dozen National Science Foundation grants and the author of many pace-setting books and articles on environmental affairs, Dr. Caldwell is presently leading efforts in Bloomington to remediate six contaminated landfills involving one of the largest superfund settlements in history.

WILLIAM COLGLAZIER is the Director of the University of Tennessee Waste Management Research and Education Institute in Knoxville, Tennessee, and is also the Director of the University of Tennessee Energy, Environment, and Resources Center. He directs research on a number of policy topics dealing with both hazardous and radioactive waste management. He has edited *The Politics of Nuclear Waste*, has chaired a task force for a Presidential commission on radioactive waste management, and is a member of the Board of Radioactive Waste Management of the National Academy of Sciences. He received a Ph.D. in Physics from the California Institute of Technology and was a research fellow at the Institute for Advanced Study, Princeton University. He served as Associate Director of the Aspen Institute Program in Science, Technology, and Humanism, and spent five years at the Kennedy School of Government at Harvard University working on science, technology, and public policy research.

MARY ENGLISH is an Assistant Director of the Energy, Environment, and Resources Center at the University of Tennessee. She has worked for state government in environmental planning and has contributed to publications in the fields of land management and environmental policy. She is currently part of a five-member research team that is analyzing value issues in radioactive waste management under a grant from the National Science Foundation. She holds a B.A. in English from Brown University and an M.S. in Regional Planning from the University of Massachusetts.

WENDY GREIDER is presently a senior official with the United States Environmental Protection Agency's Office of International Affairs, based in Washington, D.C. Representing U.S. interests within the Organization of Economic Cooperation and Development and the United Nations, Ms. Greider specializes in cross-boundary environmental problems.

DONALD HUISINGH is the Task Force Leader of Toxic Substances and Hazardous Materials Research Projects for the state of North Carolina and the leader of a UN sponsored international project on Waste Reduction and Pollution Prevention in Industries. He is a Professor of Environmental Sciences and also does consulting work on toxic and hazardous materials management technologies and policies. Dr. Huisingh has extensive experience on a wide array of environmental problems in the United States and Europe and has published *Proven Profit from Pollution Prevention* (1985) and numerous articles.

JOANNE LINNEROOTH is a researcher at the International Institute of Applied Systems Analysis in Laxenburg, Austria. She works on a variety of economic and sociological issues concerning hazardous chemicals and technologies, including siting of facilities and public acceptance. She studied industrial engineering at Carnegie-Mellon University, and earned a Ph.D. in economics from the University of Maryland. She co-authored, with H. C. Kunreuther, *Risk Analysis and Decision Processes: The Siting of Liquified Energy Gas Facilities in Four Countries* (1983).

STUART MESSINGER is presently a County Senior Planner in Queensbury, New York. As the chief staffer behind St. Lawrence County's $26 million solid waste management project, Mr. Messinger leads the county effort to draft an Environmental Impact Statement for one of New York's first mass-burn incinerators. He was a planner for the Environmental Management Council, now chaired by Bruce Piasecki.

HANS SUTTER is a Senior Research Scientist at the Umweltbundesamt (UBA) in West Berlin. As the chief official representing the Federal Republic of Germany's project on ocean incineration, Dr. Sutter is recognized as one of Europe's leading experts on hazardous waste management issues. The UBA, which has the authority to deny permit requests for ocean incineration, has also appointed Dr. Sutter as the program coordinator for the FRG's efforts to develop on-site waste reduction strategies for their most problematic chlorinated waste streams. At present, Dr. Sutter is also completing a book on the practice of waste reduction in Europe.